SCIENCEPLUS®

TECHNOLOGY AND SOCIETY

LEVEL RED

TEST GENERATOR
TEST ITEM LISTING

HOLT, RINEHART AND WINSTON

Harcourt Brace & Company

Austin • New York • Orlando • Atlanta • San Francisco • Boston • Dallas • Toronto • London

Photo/Art Credits

Abbreviated as follows: (t) top; (b) bottom; (l) left; (r) right; (c) center; (bkgd) background.

Front Cover: (tl), Martin Wendler/Okapia/Photo Researchers; (bkgd), Page Overtures; Back Cover: (bkgd), Page Overtures; (bl) Jeff Smith/PhotoSmith/Reptile Solutions of Tucson; Title Page: i (bkgd), Page Overtures; (bl) Jeff Smith/PhotoSmith/Reptile Solutions of Tucson.

All Art and Photos, unless otherwise noted, are contributed by Holt, Rinehart and Winston.

Printed in the United States of America

ISBN 0-03-095793-1 1 2 3 4 5 6 7 8 9 021 99 98 97 96

Contents

Contents, continued

Using the *SciencePlus Test Generator*

The *SciencePlus Test Generator* contains assessment items for each unit, chapter, and SourceBook unit of your textbook. For your easy reference, this *Test Item Listing* contains a complete printout of the questions and answers that come with the *SciencePlus Test Generator.* It also contains information that is specific to the installation and use of the program with *SciencePlus.*

The *User's Guide,* which is included in the *Test Generator* package, contains general directions for using the program.

Test Generator Diskettes

The *Test Generator* is delivered on a series of diskettes. The Program Disk or Disks contain the *Test Generator* program, and the Test Items Disks contain the assessment items.

Question Types

The test items are classified by question type. In the *Test Generator,* question type is called an attribute. The *Test Generator* lets you specify attributes when selecting questions for a test. It also allows you to view and to edit question attributes. Your *User's Guide* describes these features of the program.

The *SciencePlus Test Generator* has thirteen types of questions stored on the Test Items Disks. The question types are identified by a one-letter abbreviation, as shown in the chart at right.

Question Types and Codes	
A	Activity Assessment
C	Correction/Completion
D	Data for Interpretation
E	Short Essay
G	Graphic
I	Illustrative
M	Multiple Choice
N	Numerical Problem
P	Performance Task
S	Short Response
T	True/False
W	Word Usage
K	Matching

Question Difficulty Level

Some of the chapter assessment items and end-of-unit assessment items are meant to be more challenging. The *Test Generator* has an attribute for level of difficulty. Your *User's Guide* describes how to use the difficulty levels when selecting questions for a test. In *SciencePlus,* challenging questions are identified by the following one-digit level of difficulty:

3 Difficult
4 Very Difficult

Macintosh® Users

The *Test Generator User's Guide* describes how to copy the diskettes to your computer's hard drive. Look for the section on installation.

Question Disks: Folders and Files

On the Test Items disks, the files are grouped into folders, with one folder for each unit of your textbook. Each folder contains one chapter file for each chapter in the unit and four additional files. The additional files are a unit test, a SourceBook test, an activity assessment, and a bank of extra assessment items for the unit.

For example, here is a breakdown of the files in the Unit 1 folder.

Chapter.C1	Chapter 1 Assessment
Chapter.C2	Chapter 2 Assessment
Chapter.C3	Chapter 3 Assessment
Activity.A1	Activity Assessment for Unit 1
Extra.E1	Extra Assessment Items for Unit 1
Source.S1	SourceBook Assessment for Unit 1
Unit.U1	End-of-Unit Assessment for Unit 1

As the *User's Guide* explains, you can make tests by combining questions from any of the files.

Windows® Users

The *Test Generator User's Guide* describes how to install the diskettes on your computer's hard drive. Look for the section on installation.

Question Disks: Files and Directories

The test questions are stored in chapter files. After you have installed the *Test Generator* on your hard drive, the chapter files are grouped into directories, with one directory for each unit of your textbook. Each directory contains one chapter file for each chapter in the unit and four additional files. The additional files are a unit test, a SourceBook test, an activity assessment, and a bank of extra assessment items for the unit.

For example, here is a breakdown of the files in the Unit_1 directory.

Chapter.C1	Chapter 1 Assessment
Chapter.C2	Chapter 2 Assessment
Chapter.C3	Chapter 3 Assessment
Chapter.A1	Activity Assessment for Unit 1
Chapter.E1	Extra Assessment Items for Unit 1
Chapter.S1	SourceBook Assessment for Unit 1
Chapter.U1	End-of-Unit Assessment for Unit 1

As the *User's Guide* explains, you can make tests by combining questions from any of the files.

Technical Support

If you have any questions about the *Test Generator* or need assistance, call the Holt, Rinehart and Winston technical support line at 1-800-323-9239.

About *SciencePlus* Assessment

SciencePlus assessment is designed to provide you with a comprehensive set of evaluation tools. For each unit of the Pupil's Edition, you will find five categories of tests: Chapter Assessment, End-of-Unit Assessment, SourceBook Assessment, Activity Assessment, and Extra Assessment Items. The Extra Assessment Items consist of questions found only in the *Test Generator* and *Test Item Listing.* All of the other tests have blackline-master versions in the *Teaching Resources* booklets.

This *Test Item Listing* booklet provides a handy way of previewing the tests and questions contained on the *Test Generator.* The *Test Generator* is easy to use, includes graphics, and is fully customizable. You can easily change questions or add questions of your own.

Each test is self-contained and comprehensive. You will find that the Chapter Assessment and End-of-Unit Assessment each contain at least two Challenge questions to help you evaluate higher-order thinking skills. Activity Assessment, SourceBook Assessment, and Extra Assessment Items are also included for each unit.

The Activity Assessments help you assess the way students actually carry out Explorations and experiments. These assessments require students to manipulate equipment and materials and to compile and analyze data. Using the Activity Assessments, you can assess your students' abilities to solve problems using the tools, equipment, and techniques of science. Teacher's Notes are provided for each Activity Assessment to guide you in setting up the activity stations and in grading the students' efforts.

A Note About the Philosophy

In *SciencePlus* there is no clear distinction between teaching and assessing. Every suggested assessment activity, including testing, is designed to teach. Assessment should be ongoing and should measure performance in every area. The quality of student class work, homework, lab work, ScienceLog entries, and performance on tests should all be factors in assigning grades.

The authors strongly discourage reliance on recall-based assessment items. The *SciencePlus* program relies on performance-based assessment strategies. Students are provided with realistic problems to solve. Much effort has been directed toward making the assessment items consistent with the actual pursuit of scientific knowledge. Science is not a process of memorizing vocabulary; science consists of gathering data, hypothesizing, experimenting, making predictions, analyzing data, making generalizations, and many other processes. The test items in this program are designed to assess your students' mastery of these skills. In working toward this goal, the authors have used the Assessment Item Development model shown on the following page.

Assessment Item Development Model

Verbal

Word Usage: Words given are to be used in a prescribed situation.

Correction/Completion: Incorrect or incomplete sentences and paragraphs are given for correction or completion.

Short Essay: Information is given or a question is posed for short-essay response.

Short Response: Answers to these questions require a tick mark, a line, a single word, a phrase, or a sentence.

Graphic

Graph for Interpretation: A graph of a relationship between two variables is given for interpretation.

Graph for Correction or Completion: An incorrect or incomplete graph is given for correction or completion.

Graphing Data: Data are given to be graphed.

Illustrative

Illustration for Interpretation: Illustrations (drawings or photographs) are presented for interpretation.

Illustrations for Correction or Completion: An incorrect or incomplete illustration is given for correction or completion.

Answering by Illustration: A question is asked for which a drawing is the expected answer.

Numeric

Data for Interpretation: A data table is given for interpretation.

Numerical Problem: A problem is given for a numerical solution.

Selecting and Using the Assessment Items

When choosing assessment items that are appropriate for your class, consider the students' reading and writing levels, process skills, and critical-thinking skills. A brief glance at the questions, illustrations, and graphs in each test will give you a good idea of the most suitable questions for your students.

Keep in mind that the manner of testing determines what is learned: tests that require tick-mark responses teach tick-marking. Tests that require verbal, graphic, illustrative, and numeric responses develop writing, speaking, graphing, drawing, and mathematical skills. A superior test draws on as many skills as possible.

In addition to the other tests, each Activity Assessment is designed to be done in one class period. Of course, the Activity Assessments will require some advance preparation. The Teacher's Notes provide clear directions for setting up each assessment. Every effort has been made to make the Activity Assessments economical in terms of both time and materials.

Chapter 1 Assessment

Word Usage

1. Describe a stone wall in the country or in a park using the following words: environment, interaction, abiotic, and biotic.

2. Describe how clothing keeps us comfortable in the winter and the summer using the following words: tolerance, range, and temperature.

Short Response

3. Complete the chart below by putting an "X" in each row to show whether the relationship is an example of commensalism, mutualism, or parasitism.

Relationship	Commensalism	Mutualism	Parasitism
Vines live on trees without harming them.			
Humans can become infested with tapeworms.			
Barnacles attach themselves to whales as a means of ocean transportation.			
Ticks get nourishment from animals' flesh.			
Bees spread pollen as they gather nectar from flowers.			
In lichens, fungi provide moisture for algae, which manufacture food for both types of organisms.			

4. How is a niche different from a habitat?

____

____

____

Illustrative

5. Construct an interaction diagram for an environment that you are familiar with. Include four animals and four plants that would be found in that environment.

a. Suppose that dust or smoke darkened the sky for several months. What affect would this have on the plants that you have included in your diagram?

____

____

____

b. Explain how these changes would affect the population of each of the kinds of animals in this environment.

____

____

____

Data for Interpretation

6. Yolanda wanted to plant a particular plant and decided to conduct an experiment to figure out how much moisture and light to provide and what temperature was best for the plant.

   She placed 10 seeds on each of 3 petri dishes: one dry, one with a damp paper towel, and one with standing water. Then she placed 10 seeds on each of 3 more petri dishes. One of those she left in total darkness; one in constant light; and one in half darkness, half light. Finally, she placed 10 seeds on each of 3 additional petri dishes. She put one in the refrigerator (10°C), one on a table in the middle of the room (20°C), and one by the heater (30°C).

   After several days, she examined the dishes to see how many seeds had sprouted. She then graphed her results, as shown.

   a. What is this plant's range of tolerance for moisture, and what is the optimum amount needed?

   ____

   ____

   b. What kind of light does this plant need?

   ____

   ____

   c. What is this plant's range of tolerance for temperature, and what is the optimum temperature?

   ____

   ____

   d. Recommend the best conditions for growing this type of plant.

   ____

   ____

   ____

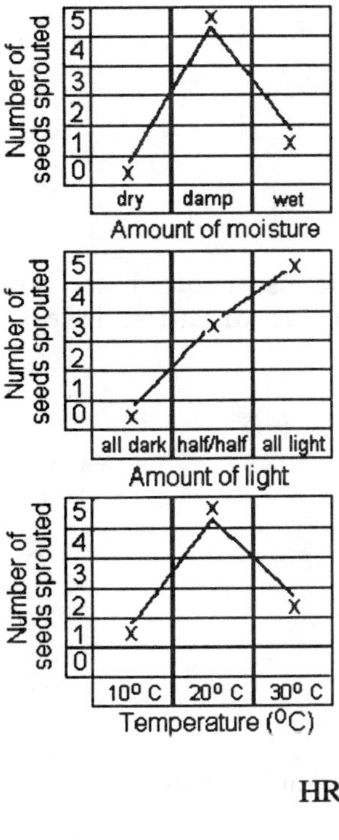

Answers to Chapter 1 Assessment

Word Usage

1. Answer:
   Sample answer: A stone wall is an *abiotic* part of the *environment*. Once vines, mosses, and other *biotic* elements make the wall their habitat, the wall becomes an excellent example of the *interaction* between the biotic and abiotic parts of the environment.

2. Answer:
   Sample answer: Wearing warm clothing in the winter and cool clothing in the summer helps us extend our *range of tolerance for temperature*.

Short Response

3. Answer:

Relationship	Commensalism	Mutualism	Parasitism
Vines live on trees without harming them.	X		
Humans can become infested with tapeworms.			X
Barnacles attach themselves to whales as a means of ocean transportation.	X		
Ticks get nourishment from animals' flesh.			X
Bees spread pollen as they gather nectar from flowers.		X	
In lichens, fungi provide moisture for algae, which manufacture food for both types of organisms.		X	

4. Answer:
   Sample answer: A niche is the way of life that an organism adopts in order to survive in a particular habitat. A habitat is simply the location where an organism lives.

Illustrative

5. Answer:
   Answers will vary, but plants and animals should be logical choices for a shared environment, and arrows should correctly and completely indicate producer-consumer relationships.

   a. The plants would not be able to produce food because they would not receive any sunlight.
   b. The animals that ate plants would die of starvation if there were no plants, and the animals that ate other animals would also die of starvation if there were no plant-eating animals.

   Difficulty: 3

Data for Interpretation

6. Answer:
   a. Damp to wet; damp
   b. This plant will grow best if given as much light as possible.
   c. Approximately 10°C to 30°C; 20°C.
   d. This plant will grow best in damp, well-lit conditions where the temperature stays at 20°C.

   Difficulty: 4

Chapter 2 Assessment

Word Usage

1. Use the words *herbivore, decomposer, producer,* and *scavenger* to describe the following food chain.

   shrub ⟶ deer ⟶ fox ⟶ vulture ⟶ bacteria

   ____

   ____

   ____

   ____

   ____

   ____

2. Use the words *coniferous* and *deciduous* in one or two sentences to show that you know what they mean.

   ____

   ____

   ____

   ____

   ____

Short Response

3. Fill in the blank to match each term on the left with its example on the right.

   a. producers ___ Minnows eat algae.

   b. primary consumers ___ Humans catch and eat fish.

   c. secondary consumers ___ Oak trees grow in sunlight.

   d. top carnivores ___ Frogs eat grasshoppers but avoid snakes.

Illustrative

4. Study this illustration and show what you know about interactions between living things by completing the following tasks.

   a. Use the illustration to name at least one example of each of the following: consumer, producer, scavenger, predator and its prey, decomposer.

   ____

   ____

   ____

   ____

   ____

   ____

   b. Use the illustration to describe examples of the following:
   - herbivore

   ____

   ____

   ____

   - carnivore

   ____

   ____

   - omnivore

   ____

   ____

c. Some of the interactions taking place in the setting may not be visible. Describe a possible example.

d. Describe the tolerance of a plant to extremes of temperature, light, or moisture.

e. Describe the tolerance of an animal to extremes of temperature or moisture.

f. Describe an interaction between an abiotic and a biotic factor of the environment that you haven't already described.

Short Essay

5. Using what you've learned about food chains, food webs, and interactions in the environment, support or refute the following statement:

> "We can never know how wide a circle of disturbance we produce in the harmonies of nature when we throw the smallest pebble into the ocean of organic life."
>
> George Perkins Marsh

Answers to Chapter 2 Assessment

Word Usage

1. Answer:
   Sample answer: The shrub is the *producer*, providing energy for the deer, a *herbivore*. The deer is eaten by a *carnivore*, the fox. When the fox dies, it is eaten by the vulture, a *scavenger*. The remains of all the organisms in the chain become food for bacteria, which are *decomposers*.

2. Answer:
   A *coniferous* tree is a tree that has cones instead of flowers and needle-shaped leaves that do not fall from the tree in the fall. A *deciduous* tree has flowers and flat leaves that are dropped in the fall.

Short Response

3. Answer:
   <u>b</u> Minnows eat algae.
   <u>d</u> Humans catch and eat fish.
   <u>a</u> Oak trees grow in sunlight.
   <u>c</u> Frogs eat grasshoppers but avoid snakes.

Illustrative

4. Answer: a. Consumers: bird, squirrel, rabbit, deer, bobcat, vultures, snake, frog. Producers grass, trees, bushes. Scavenger: vulture. A predator and its prey: bobcat and dee snake and frog, bird and worm. Decomposers: fungi such as mushrooms, bacteria (although too small to be visible in the illustration).
   b. • herbivore
   Sample answer: A rabbit is a *herbivore* because its main food is plants.
   • carnivore
   Sample answer: A bobcat is a *carnivore* because its main food is animals.
   • omnivore
   Sample answer: A bird is an *omnivore* because it comsumes both insects and seeds.
   c. Acceptable answers might describe an interaction that is hidden from view, such a one taking place in the soil or the interaction of microscopic organisms such as bacteria with their environment.
   d. Examples include the following: The plants growing under trees must be able to tolerate conditions in which the amount of light is limited. Grass can tolerate a wide temperature range, including frosts.

Answer:

e. Except perhaps for the birds, the animals in the illustration are well-adapted to survive in both summer and winter temperatures. Their activity in winter, however, is much reduced. Snakes and frogs are dormant where winter temperatures are below freezing. None of the animals can survive indefinitely without water.

f. Examples include descriptions of the interaction of any of the animals or plants shown with air, soil, or light.

Difficulty: 3

Short Essay

5. Answer:

Accept all reasonable responses. Sample answer: Based on what we know about food chains, food webs, and the flow of energy within a community, it is clear that the roles within the natural world are so interconnected that even the smallest change can affect entire communities. Like the ripples that are caused by a pebble tossed into an ocean that travel outward in circles, the repercussions from a tiny disturbance within the environment can cause a ripple effect among a wide range of interacting organisms and their surroundings.

Difficulty: 4

Chapter 3 Assessment

Word Usage

1. Use all of the terms below in one or two sentences to show how the terms are related.
   a. needs, future, sustainable development, planning, resources

   ____

   ____

   ____

   ____

   b. organisms, time, succession, area

   ____

   ____

   ____

   ____

Short Response

2. For each substance listed in the table below, place an "X" in the proper column to show whether it is an acid or a base.

Substance	Acid	Base
orange juice		
baking soda		
ammonia		
vinegar		
crushed limestone		

Numerical Problem

3. When pH drops by one whole point, acid content increases 10 times. Last year, the average pH of a lake in Pennsylvania was 7.0. This year, the average pH has dropped to 5.0.
   a. By how many times did the acid content increase? Show your work.

   b. Why is it important to record this kind of data on a yearly basis?

Graphic

4. Examine the map and then answer the questions that follow.
   a. Where does the worst acid rain seem to be concentrated? Why?

   ____

   ____

   ____

   ____

   ____

   b. What might explain the strongly acidic rainfall in Birmingham? (It is not a large city.)

   ____

   ____

   ____

   ____

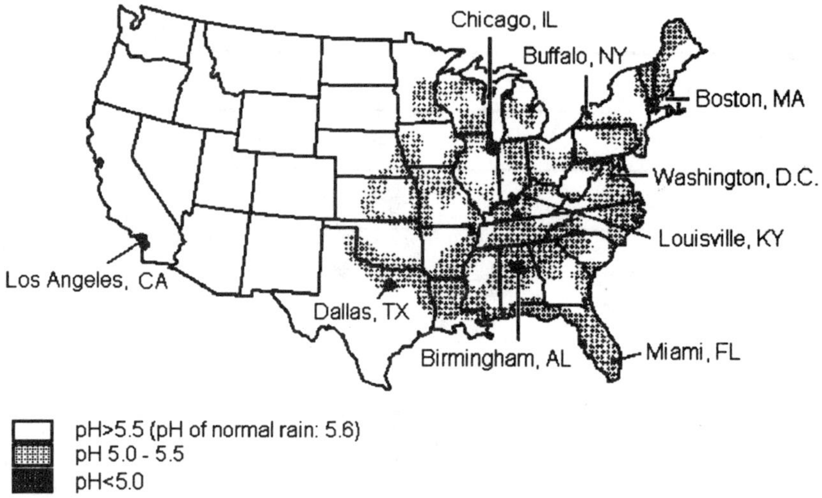

Short Essay

5. Over the past 25 years, the pollution produced by automobiles has declined. Now, the average late-model car produces a fraction of the pollutants (except carbon dioxide) produced by the average 1970-model car. Given this trend, do you expect automobile pollution to decline to zero in the near future? Explain your answer.

Answers to Chapter 3 Assessment

Word Usage

1. Answer:
   a. *Sustainable development* involves careful *planning* in terms of using our natural *resources*, so that we can provide for the *needs* of *future* generations.
   b. New *organisms* constantly replace old, lost, or destroyed organisms in a certain *area* over long periods of *time*. This process is called *succession*.

Short Response

2. Answer:

Substance	Acid	Base
orange juice	X	
baking soda		X
ammonia		X
vinegar	X	
crushed limestone		X

Numerical Problem

3. Answer:
   a. A drop from pH 7.0 to pH 6.0 increased the acidity by 10 times, and then a drop from pH 6.0 to pH 5.0 increased the acidity by another 10 times. 10 X 10 = 100 times as much acid content
   b. Yearly records of acid content in lakes provide evidence of increases or decreases in acid rain. These figures may help scientists determine what factors contribute to the formation of acid rain, as well as precautions that should be taken by industries producing substances that increase the amount of acid rain in a certain region. The information may be compared with known output of waste products from industrial sources and later used by local or national governments in efforts to reduce the problems caused by acid rain.

Graphic

4. Answer:
   a. The worst acid rain is most concentrated near major metropolitan areas where large numbers of people, industrial machinery, and automobiles contribute to the pollution that causes acid rain.
   b. Since Birmingham is not a major metropolis, its acidic rainfall must be caused by a major industry located nearby.

Difficulty: 3

Short Essay

5. Answer:
   Student answers will vary, but students should recognize that while reductions in the total amount of pollutants produced may continue to decline, pollutants cannot be expected to decline to zero because car engines will always produce some pollutants. In addition, because there are increasing numbers of automobiles on the road each day, the gain in improvement is partially lost.

   Difficulty: 4

Unit 1 End-of-Unit Assessment

Word Usage

1. Use the words *change* and *time* in a sentence that describes the process of *succession*.

2. Use all of the following words in one or two sentences to show how they are related: bacteria, mold, consume, mushrooms, environment, and decomposer.

Correction/Completion

3. The statement below is either incorrect or incomplete. Your challenge is to make it correct and complete.

   Abiotic factors have no effect on the interactions that take place among the biotic factors of an environment.

Short Response

4. List as many abiotic parts of the following communities as you can.
   a. desert

   ____

   ____

   ____

   b. rain forest

   ____

   ____

   ____

5. Give an example of each of the following relationships:
   a. mutualism

   ____

   ____

   ____

   b. commensalism

   ____

   ____

   ____

   c. parasitism

   ____

   ____

   ____

Data for Interpretation

6. Jamal placed three groups (A, B, C) of 10 radish seeds between layers of soaked paper towels. Then he placed each group in a different place. After 4 days, he found that in one sample, eight seeds had sprouted. In another, four seeds had sprouted. In the third sample, no seeds had sprouted. Fill in the chart and match each of the samples with Jamal's results.

Group	Place	Number of seeds sprouted
A	warm shelf	
B	cool shelf	
C	refrigerator	

Explain Jamal's results.

Short Response

7. Briefly describe the habitat and niche of the following organisms:
   a. shark

   b. caterpillar

Data for Interpretation

8. The table below shows some of the living things in a salt marsh and the foods they depend on.

Organism	Food/energy source
marsh grass	sun
crabs	algae and particles of dead matter
grasshoppers	marsh grass
snails	algae, marsh grass, dead animals, particles of dead matter
spiders	grasshoppers, plant hoppers, insects
raccoons	crabs, mussels, snails
dragonflies	grasshoppers, plant hoppers
mussels	algae
algae	sun

a. Fill in the table below with the appropriate organisms from the table above.

Producers	Herbivores	Carnivores	Scavengers

b. Using the first table, make a food web with as many lines as possible. Make sure the head of the connecting arrow is pointed *away* from the organism that constitutes the food.

Short Essay

9. In what ways are producers and decomposers opposites of each other?

Graphic

10. You have learned that burning oil and coal contributes to pollution that causes acid rain. Burning fossil fuels also releases large quantities of carbon dioxide into the atmosphere. Below is a graph that relates carbon dioxide levels to atmospheric temperature.

    a. What can you infer from the graph?

    ____

    ____

    ____

    ____

    ____

    b. The theory that deals with this relationship is called *global warming.* What other kinds of human interactions with the environment could contribute to global warming?

    ____

    ____

    ____

    ____

    ____

    c. What kinds of things can you do to help stop global warming?

    ____

    ____

    ____

    ____

    ____

Amount of Carbon Dioxide in the Atmosphere

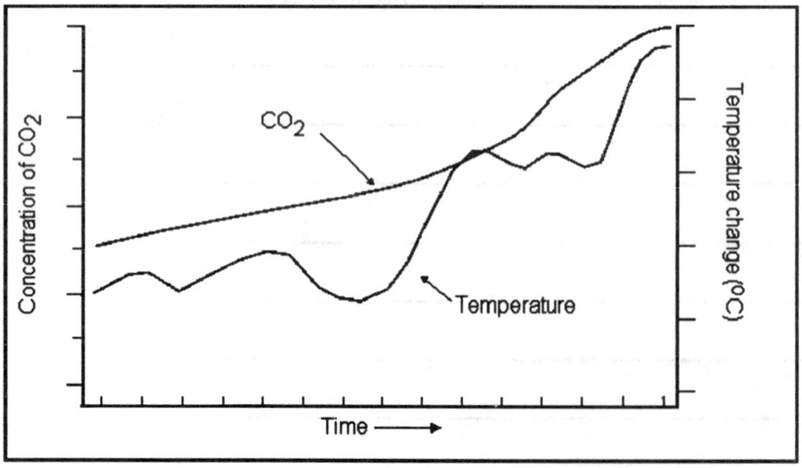

Short Essay

11. A soybean farmer in Illinois notices that his bean plants are being destroyed by moth larvae that seem to have no natural predators. He is considering importing a lizard from Cambodia that is a natural predator of this species of moth. What do you think of this idea?

12. How might a switch from fossil fuels (oil, coal, and gas) to alternative forms of energy affect the average adult life expectancy?

Answers to Unit 1 End-of-Unit Assessment

Word Usage

1. Answer:
   Sample answer: *Succession* is the process of *change* over *time* as living organisms replace one another in a natural environment.

2. Answer:
   Mushrooms, *mold*, and *bacteria* all play a common role in the *environment*. Each is a *decomposer* that breaks down and *consumes* dead material.

Correction/Completion

3. Answer:
   Sample answer: Abiotic factors *help to determine* the interactions that take place among the biotic factors of an environment. The *biotic and abiotic factors are intricately connected to each other.*

Short Response

4. Answer:
   a. Sample answer: Sand, sun, minimal water, wind, high daytime temperature, low nighttime temperature
   b. Sample answer: Rain, tropical temperature, high humidity, nutrient-rich soil, filtered sunlight

5. Answer:
   a. Sample answer: In lichen, the fungus provides protection for the algae, which supplies the food for itself and for the fungus.
   b. Orchids grow on other plants for mechanical support, without harming the other plants.
   c. Cuckoos replace other birds' eggs with their own so that the other birds incubate the cuckoos' eggs.

Data for Interpretation

6. Answer:
   No seeds sprouted in group C because a refrigerator's temperature is below the range of temperatures in which radish seeds will sprout. Only a few seeds sprouted in group B because the cool shelf, although within the range of tolerance for temperature for the seeds, was not warm enough for the best sprouting conditions. The warm shelf gave the optimum temperature, as shown by the number of seeds sprouted in group A.

Group	Place	Number of seeds sprouted
A	warm shelf	8
B	cool shelf	4
C	refrigerator	0

Short Response

7. Answer:
   a. The natural habitat of a shark is an ocean. Its niche is defined by its lifestyle: it eats fish and dead or decaying organisms, and it is eaten by other sharks and large marine animals.
   b. A caterpillar's habitat consists of leafy plants and trees. Its niche is defined by the fact that it eats the leaves of plants and is eaten by other insects and birds.

Data for Interpretation

8. Answer:
   a.

Producers	Herbivores	Carnivores	Scavengers
marsh grass algae	crabs grasshoppers mussels	spiders raccoons dragonflies crabs	crabs snails

   b. Sample answer:

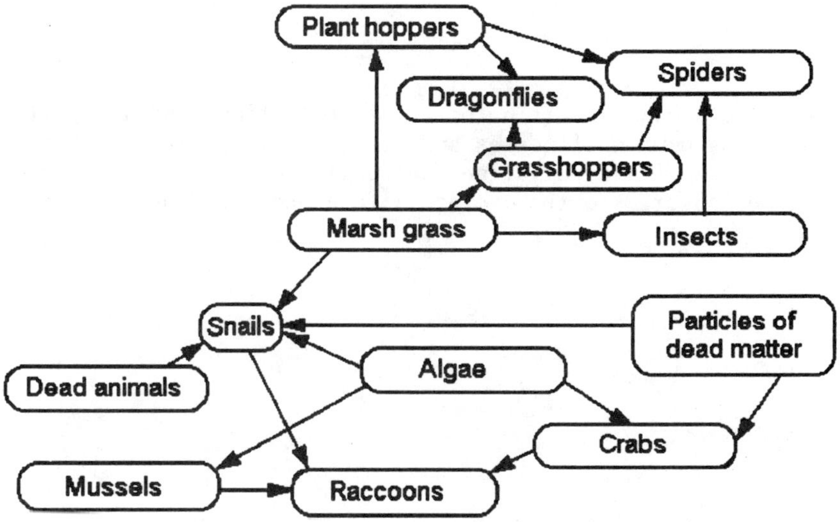

Short Essay

9. Answer:
Sample answer: Producers make new food from raw materials for themselves as well as for consumers. They represent the beginning of a food chain or food web. Decomposers make their food by breaking down dead organisms, both consumers and producers, into raw materials. They represent the end of a food chain or food web.

Difficulty: 3

Graphic

10. Answer:
a. There is a direct relationship between increasing levels of carbon dioxide and increasing atmospheric temperature. The graph seems to indicate that as more carbon dioxide is emitted, the temperature rises.
b. Answers will vary, but students should recognize that any human activity that affects carbon dioxide levels may contribute to global warming. For example, deforestation could contribute to global warming because trees remove carbon dioxide from the atmosphere. Burning forests to create farmland could also contribute to the process because the burning of trees adds carbon dioxide to the atmosphere.
c. Sample suggestions: Start an awareness program at school, encourage the planting of new trees, learn about carpool programs, and support local and national government leaders who encourage environmental awareness.

Difficulty: 3

Short Essay

11. Answer:
Sample answer: If the lizards are natural predators of the moth species, they will kill all of the larvae. However, if the lizards have no natural predators, which is possible because their natural habitat is in Cambodia, there will be a population explosion of the lizards. Then the farmer will have another problem. Another possible solution for the farmer might be to develop a strain of soybeans that is not preyed on by the moth larvae. In this way, the farmer would not upset this ecosystem.

Difficulty: 4

12. Answer:
Sample answer: Fossil fuels are a major contributor to air pollution, and air pollution has been known to aggravate many common illnesses and diseases. Thus, if alternative forms of energy reduce the amount of air pollution, the health of many people would improve. They would then live longer, and the average life expectancy would increase.

Difficulty: 4

Unit 1 Activity Assessment

Activity Assessment

1. **Community Interaction**

Teacher's Notes

Overview
Students build a terrarium and make observations about interactions among its components. Then students create an interaction diagram.

Safety Alert! Remind students to exercise caution when transporting the earthworm to its new environment.

Materials
(per activity station)
- a terrarium made from a 2 L plastic bottle (see diagram) or a glass bowl
- 0.5 L of soil
- a beaker of water (about 100 mL)
- 2 small, hardy plants (such as pothos ivy, Swedish ivy, or coleus)
- large and small rocks
- an earthworm
- a light source (sunlight or bright lamp)

Preparation
Prior to the assessment, equip student activity stations with the materials needed for each experiment.

Time Required
Each student should have 25 minutes at the activity station and 15 minutes to complete the Data Chart.

Performance
At the end of the assessment, students should turn in the following:
- a terrarium
- a completed Data Chart

Evaluation
The following is a recommended breakdown for evaluation of this Activity Assessment:
- 25% ability to create an adequate environment for the earthworm and plants
- 25% ability to identify biotic and abiotic factors within the community
- 50% accurate diagram of the interactions in the terrarium

Safety Alert!

Plastic-Bottle Terrarium

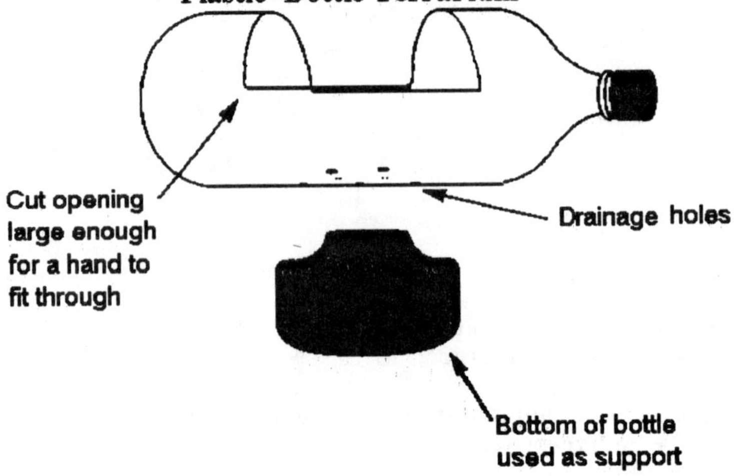

2. **Community Interaction**

All things in an environment interact with each other. Each environment contains parts that are living, once-living, and nonliving. At your activity station, you will find materials to create a terrarium. In the terrarium you will use biotic and abiotic elements to create an environment for an interactive community. When you're done, you'll be able to see interactions in a community firsthand!

Before You Begin . . .

As you work through the tasks, keep in mind that your teacher will be observing

- how you use the materials to create an adequate environment for the earthworm and plants
- how well you identify the biotic and abiotic factors within the environment
- how clearly you diagram the interactions in the terrarium

Get Ready to Interact!

Task 1 Examine the materials at your activity station. Decide how to construct the best environment for the earthworm and the plants. Then put the items together, remembering to use caution when handling living things.

Task 2 Identify and observe the biotic and abiotic factors in your community. Record the biotic and abiotic factors and the interactions in the Data Chart. There will be some interactions that you may not be able to see, but can infer.

Task 3 On a separate sheet of paper, draw a diagram that connects all of the interacting factors in your terrarium. Then summarize your observations.

Safety Alert!

3. **Data Chart**

	Observations	Interactions
Biotic Factors		
Abiotic Factors		

Answers to Unit 1 Activity Assessment

1. Answer: Not applicable (teacher's notes)

2. Answer: Not applicable (student's notes)

3. Answer:
   Answers will vary depending on the terrariums. However, students should demonstrate the ability to create an adequate environment for the earthworm and plants, correctly identify the biotic and abiotic factors within the community, and accurately diagram the interactions taking place in the terrarium.

Unit 1 SourceBook Assessment

Word Usage

1. Place the following terms in order from the most general term to the most specific: ecosystems, reproductive groups, communities, biosphere, and populations.

____

____

____

Short Response

2. Mr. Jones owned the neighborhood pet-food shop. One day Tom came in with his dog, Fido, and Susie dropped by with her parakeet, Tweety. They were both looking for food for their pets. From this story, how many different populations were in the store?

____

____

____

3. Put an "X" by each of the following that represents an ecosystem.

____ the Atlantic Ocean
____ your backyard
____ a fish tank
____ New York City
____ your schoolroom
____ a leaf

Illustrative

4. Draw a food pyramid using the following organisms: snake, grass, eagle, and rabbit.

Short Essay

5. Describe one way that human activity could lead to the extinction of a species.

6. Name one type of pollution that is common in the United States and one cause of that pollution.

7. In some communities, the garbage is disposed of in landfills, where a layer of clay is placed over the garbage. Other communities dispose of solid wastes in open dumps. Which type of solid-waste management is better for humans and for the environment, and why?

8. Explain an animal's role in the nitrogen cycle.

____

____

____

____

____

____

Correction/Completion

9. The community that exists at the final stage of primary succession is called the _______________.

True/False

10. Limiting factors determine where a species can live and how large its population can grow.
    a. true b. false

11. The process of secondary succession takes longer than the process of primary succession.
    a. true b. false

12. Plankton have relatively large fins that enable them to swim to great depths in the marine biome.
    a. true b. false

13. Spraying chemicals on lawns to kill weeds can lead to
    •water pollution. a. true b. false
    •soil pollution. a. true b. false

Multiple Choice

14. The sequoia trees in California take in carbon dioxide from the environment to make food for themselves and give off _________________ as a waste product.
    a. oxygen b. carbon dioxide c. nitrogen d. hydrogen

15. Most of the Earth's atmosphere consists of
    a. oxygen. b. carbon dioxide. c. nitrogen. d. hydrogen.

16. Geographic areas of the Earth that share similar climates and can be identified by their plant life and soil type are called
    a. ecosystems. b. biomes. c. hemispheres. d. continents.

17. Which of the following is NOT an example of a pioneer species in primary succession?
    a. moss b. grass c. lichens

18. Which of the Earth's terrestrial biomes is located just south of the Arctic icecap and is characterized by freezing temperatures and little sunlight?
a. desert b. tundra c. tropical rain forest d. grassland
e. coniferous forest f. deciduous forest

19. Which biome supports the greatest variety of life found anywhere on Earth?
a. desert b. tundra c. tropical rain forest d. grassland
e. coniferous forest f. deciduous forest

20. Iowa cornfields are one example of this terrestrial biome.
a. desert b. tundra c. tropical rain forest d. grassland
e. coniferous forest f. deciduous forest

21. Christmas trees, such as spruces or pines, are members of this biome.
a. desert b. tundra c. tropical rain forest d. grassland
e. coniferous forest f. deciduous forest

22. This part of the marine biome has no light and some very strange creatures.
a. oceanic zone b. intertidal zone c. neritic zone
d. estuaric zone

23. Which freshwater biome supports more life?
a. a lake b. a swift-moving stream

24. Which of the following is NOT an example of a renewable resource?
a. trees b. soil c. fish d. water

25. With respect to the study of conservation, why is it important that all hunters have a hunting license?
a. to make their sport legal
b. to prevent them from shooting another hunter by accident
c. to limit the age, size, sex, and number of prey that each hunter kills
d. to raise money for the state by requiring a fee for the license

Answers to Unit 1 SourceBook Assessment

Word Usage

1. Answer: Biosphere, ecosystems, communities, populations, reproductive groups

Short Response

2. Answer:
   There were three populations in the store: the populations of humans, dogs, and parakeets.

3. Answer:
   X the Atlantic Ocean
   X your backyard
   X a fish tank
   X New York City
   X your schoolroom
   X a leaf

Illustrative

4. Answer:

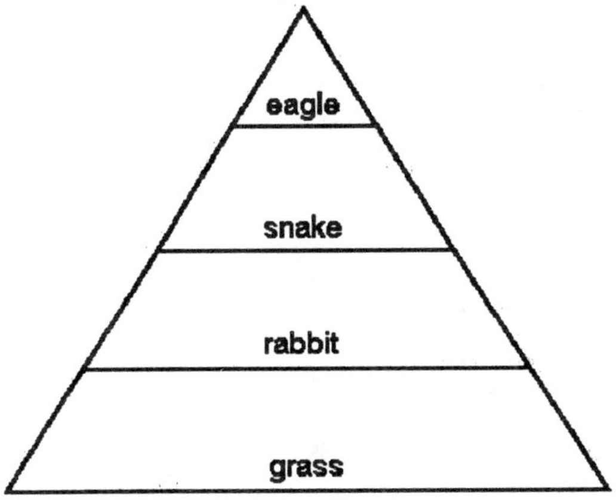

Short Essay

5. Answer:
   Answers may include the following ideas: Deforestation destroys the homes of ground animals and birds. Water polluted with chemical wastes kills fish and other aquatic life and may kill plant life if used for irrigation. Forest fires kill trees and animals and destroy organisms' homes. Air pollution affects plant and animal respiration. Overhunting may drive a species to extinction.

6. Answer:
Possible answers include the following: Air pollution is caused by car exhaust and industrial smokestacks; water pollution results from chemical dumping and runoff from landfills; soil pollution is caused by chemicals put on lawns and by acid rain; and noise pollution results from loud machines and traffic.

7. Answer:
Landfills are better./ Explanations may vary. Sample explanation: The layer of clay over the garbage keeps out disease-carrying rodents and helps keep rainwater from washing through the garbage and contaminating our water supplies.

8. Answer:
Animals obtain nitrogen by eating plants that absorb nitrogen through their roots with the help of nitrogen-fixing bacteria. The animals return nitrogen to the environment in the form of waste products. Bacteria break down this animal waste and the nitrogen goes back to the atmosphere as a gas.

Correction/Completion

9. Answer: climax community

True/False

10. Answer: a. true

11. Answer: b. false

12. Answer: b. false

13. Answer:
•water pollution. a. true
•soil pollution. a. true

Multiple Choice

14. Answer: a. oxygen

15. Answer: c. nitrogen.

16. Answer: b. biomes.

17. Answer: b. grass

18. Answer: b. tundra

19. Answer: c. tropical rain forest

20. Answer: d. grassland

21. Answer: e. coniferous forest

22. Answer: a. oceanic zone

23. Answer: a. a lake

24. Answer: b. soil

25. Answer: c. to limit the age, size, sex, and number of prey that each hunter kills

Unit 1 Extra Assessment Items

Word Usage

1. Use the words *scavenger* and *decomposer* in a sentence together to show that you understand their meaning.

2. Explain the difference between the following pairs of words.
   a. biotic factors and abiotic factors

   b. herbivore and carnivore

   c. producer and consumer

   d. predator and prey

Correction/Completion

3. A _________ is the way of life that a plant or animal adopts to survive in a particular _________.

4. Animals that eat dead animals or dead organic material are called _________.

5. Mushrooms, molds, and bacteria are examples of _________.

6. The following were given by students as examples of a certain food chain. Only one is completely correct. Identify the correct chain, and beneath each of the others, explain why they are incorrect.
   a. Chloe: minnow (small fish) → elodea (small water plant) → beetle → frog → human

   ____

   ____

   ____

   b. C.J.: elodea → mosquito larva → frog → bass → human

   ____

   ____

   ____

   c. Austin: elodea → turtle → minnow → bass → human

   ____

   ____

   ____

Short Response

7. After each sentence below, indicate whether the sentence describes an organism's habitat or its niche.
   a. A fly is caught in a web and eaten there. _________

   b. Mushrooms are found growing on rotting logs. _________

   c. Fleas live in an animal's fur. _________

   d. Beavers make a pond by damming a stream. _________

8. In each case below, identify whether the relationship between the two organisms is an example of commensalism, parasitism, or mutualism, and explain why.

a. A heartworm lives around the heart of a dog, weakening it and making it more vulnerable to diseases.

____

____

____

b. A rhinoceros allows an egret (a white bird) to stand on its back and eat infectious bugs.

____

____

____

c. A vulture watches a lion until it kills something. When the lion is done eating, the vulture eats what is left.

____

____

____

9. Darius likes to add the following to his hamburger: cheese, lettuce, tomato, dill pickle, bacon, egg, onions, and ketchup.
Separate all the foods in the hamburger into two groups:

a. those that came from plants

b. those that came from animals

c. For each kind of food that came from animals, what does the animal eat?

d. Now link all of the separate foods in the hamburger back to their ultimate source of energy—the sun.

10. Correctly classify the pairs of organisms according to the categories given below.

- honeybee and flower
- mistletoe and oak tree
- algae and coral reef
- tapeworm and child
- barnacles and crab
- suckerfish and shark

Mutualism	**Commensalism**	**Parasitism**

11. Answer the following questions about a polar community:

a. Identify as many abiotic parts of a polar community as you can.

____

____

____

b. Choose one type of plant or animal species that is found in a polar environment, and describe how it has adapted to the abiotic factors in that environment.

____

____

____

____

____

12. Using complete sentences, describe an example that shows you know the scientific meaning of each of the following words. For instance, your example for one of the words below might be "The diet of deer consists only of plants, such as grass, twigs, and fruit. They do not eat other animals."

a. carnivore

____

____

____

b. herbivore

____

____

____

c. omnivore

____

____

____

d. scavenger

____

____

____

e. decomposer

____

____

____

13. Select an animal and describe its interaction with the environment. In your description, mention how this animal uses other things in the environment and how it is used by other things. Use the words *habitat* and *niche* in your description.

Choice of animal: ________________________

Description of its interaction with its environment:

____

____

____

____

____

14. Select a plant and describe its interaction with the environment. In your description, mention how this plant uses other things in the environment and how it is used by other living and nonliving things. Use the words *habitat* and *niche* in your description.

Choice of plant: ________________________

Description of its interaction with its environment:

____

____

____

____

____

Graphic

15. The graph shows the amount of grass produced during a 13-week period. (For example, the number 2 on the horizontal axis indicates the end of the second week.) Suppose that mice are the only consumers of this grass and that the mice and grass are not threatened by drought, parasites, disease, or predators. Answer the following questions based on the graph:

a. At the end of which week is there the greatest amount of grass?

____

b. At the end of which week is there the least amount of grass?

____

c. Predict how the mouse population will change as a result of changes in the amount of grass available.

____

____

____

____

____

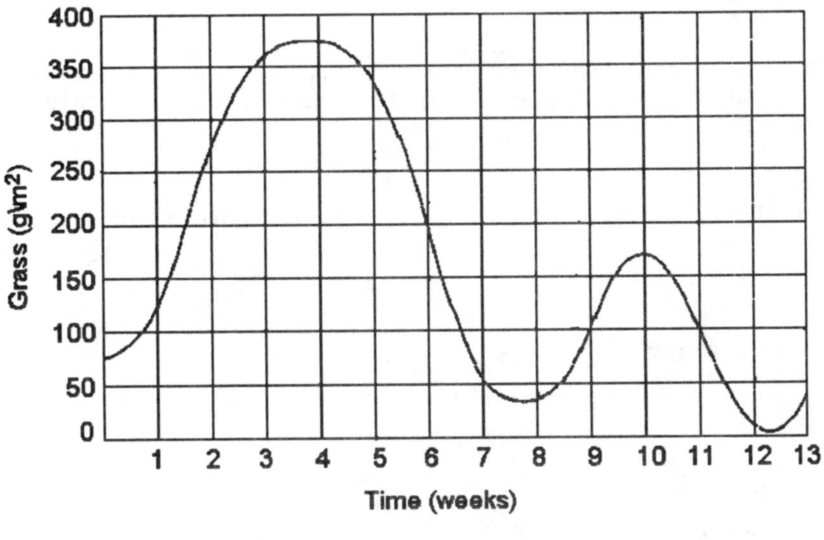

16. Study the graph of frog embryo development at different temperatures. Using the terms *range of tolerance for temperature* and *optimum temperature*, describe the information presented in the graph.

____

____

____

____

____

Development of Frog Embryos at Different Temperatures

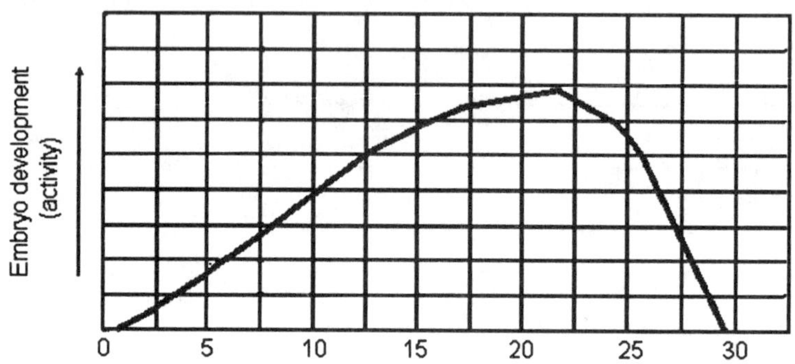

17. This graph shows the populations of bass and perch in a pond. Both types of fish are preyed on by larger types of fish. Label the graph with the letters a–d to indicate where the following situations occur:
   a. the original population of bass
   b. the largest population of perch
   c. the bass population after the larger species of fish have preyed more on the bass than on the perch
   d. the return to nearly original population counts of bass and perch

Population Changes

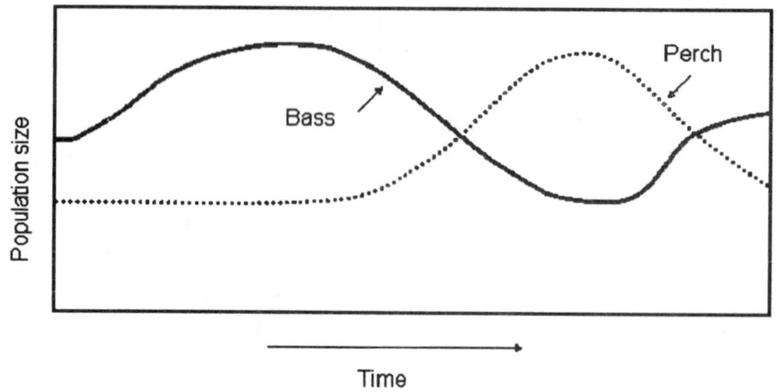

Illustrative

18. Use the diagram to answer the following questions:
    a. What do rabbits eat?

    ____

    b. Which living things shown here are producers?

    ____

    ____

    c. Which living things shown here are consumers?

    ____

    ____

    d. Which animals shown here have no predators?

    ____

    ____

    e. Imagine that the diagram shows every single living thing in this community. What would happen to the other members of the community if all of the lettuce died?

    ____

    ____

    ____

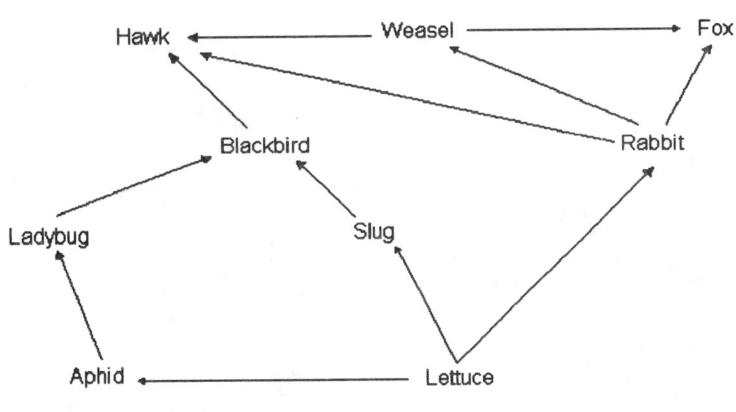

A → B = A is eaten by B

19. State the relationship between the two things shown in each picture below (for example, consumer-producer, predator-prey, or decomposer-food).

a. ______________________________

b. ______________________________

c. ______________________________

d. ______________________________

e. ______________________________

f. ______________________________

g. ______________________________

h. ______________________________

20. The food web shows some interrelationships between living things. Using the web, answer the questions that follow.

a. Identify each of the following as producers, herbivores, carnivores, scavengers, or decomposers:

mice ______________

grasses ______________

coyotes ______________

owls ______________

bacteria ______________

b. Do grasses feed on bacteria? Explain what is meant by the arrows labeled *bacteria*.

____

____

____

____

____

____

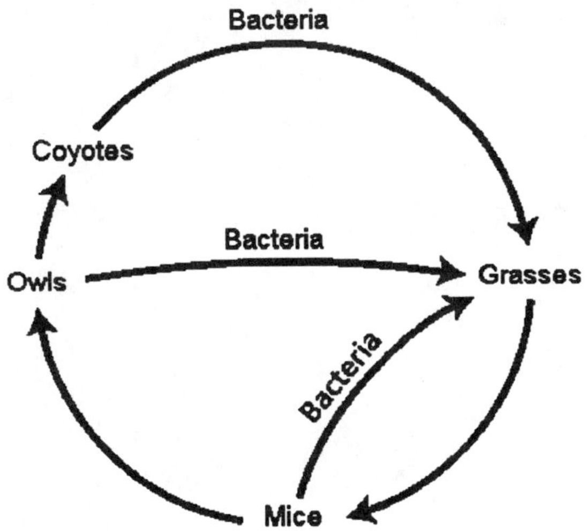

21. a. In each box in the food web, write either *producer*, *omnivore*, *carnivore*, or *herbivore*.

b. Use the words *predator* and *prey* to describe the animals in the web.

____

____

____

c. What would probably happen to the animals in this web if the number of blackberry bushes increased?

____

____

____

d. What are some other organisms that would exist in this food web if it were complete?

____

____

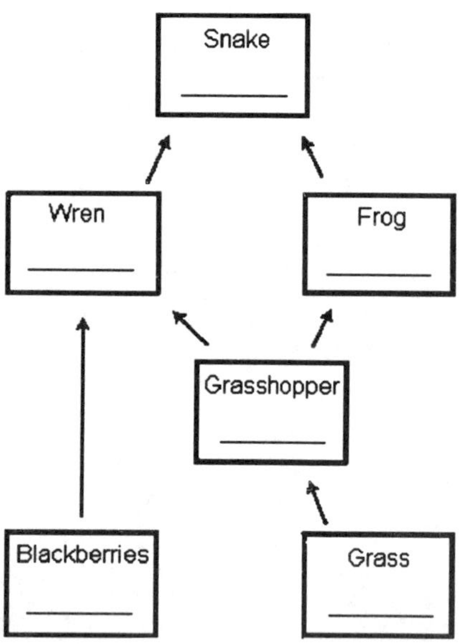

22. Connect the elements with lines to show how they interact. You should indicate 10 possible interactions. Then, in the space below, briefly describe the nature of those interactions. Make sure that you include abiotic-abiotic, biotic-biotic, and abiotic-biotic interactions.

a. ____________________

b. ____________________

c. ____________________

d. ____________________

e. ____________________

f. ____________________

g. ____________________

h. ____________________

i. ____________________

j. ____________________

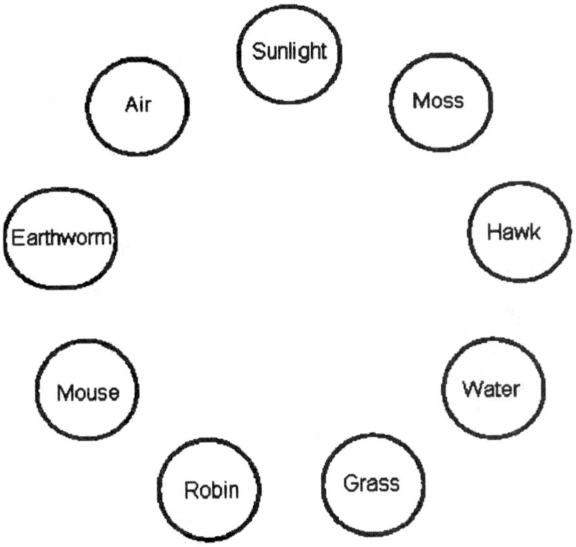

23. The pictures show the process of succession. Label the pictures *A–D* to show the correct order of succession.

24. Create an illustration showing one of the following types of interactions: consumer-producer, biotic-abiotic, predator-prey. Then name the type of interaction and use a complete sentence to describe the relationship shown.

a. Type of relationship: ______________________________

b. Description:

____

____

____

Animal	Food
herring	plankton, seaweed
mackerel	herring, seaweed
sand lance	plankton
salmon	sand lance, mackerel
whale	mackerel, salmon

Plankton are tiny floating green plants. Seaweed are green plants.
The sand lance, herring, mackerel, and salmon are all types of fish.

25. Use the chart above to complete the following tasks:
    a. Construct a food web showing the relationships between the organisms in the chart.
    b. Which organisms are

    producers? ______________________________

    ____

    consumers? ______________________________

    ____

    herbivores? ______________________________

    ____

    carnivores? ______________________________

    ____

    omnivores? ______________________________

    ____

    c. Name two pairs of organisms that are in competition for food in this environment. Also indicate the food for which the pair is competing.

    ____

    ____

    ____

    ____

    d. If a disease were killing the mackerel, what effects would this have on each of the other organisms in the web?

    ____

    ____

    ____

    ____

    ____

    ____

e. In this environment, sculpins and skates perform the role of scavengers. What would they do as the disease killed off the mackerel?

____

____

____

f. Which organisms in this environment turn the sun's energy into food for the other organisms?

____

g. Describe the niche of the whale. Include abiotic factors as well as biotic factors.

____

____

____

____

____

____

____

Performance Task

26. You are the owner of an environmental consulting firm that has been asked to do an environmental assessment of your local area. You know that there is growing concern in the community about the state of the environment and the possible effect of declining environmental conditions on humans.

    Your task is to identify (a) the major biotic and abiotic components of the environment for your area, (b) the interactions between these components, and (c) any risks to the environment posed by current human activity or natural occurrences. You have also been asked to make recommendations about what actions could be taken to protect the state of your environment.

    Task 1: Use field notes, sketches, photographs, and videotape to record your observations of this environment. Be sure to use the terms and concepts that you've learned in your study of this unit.

    Task 2: Prepare a short written report for your client. The report should describe your procedures, results, and recommendations.

    Task 3: Prepare a brief oral presentation to communicate your findings and recommendations to your client (the class). Support your presentation with overhead transparencies, posters, photographs, videotape, or any other aid that you think will help convey your results and recommendations in an effective manner.

Answers to Unit 1 Extra Assessment Items

Word Usage

1. Answer:
   Sample answer: *Scavengers*, such as vultures, ravens, eagles, and lions, feed on dead animals. *Decomposers*, such as mushrooms and bacteria, break down dead and decaying plants and animals, returning nutrients to the soil.

2. Answer:
   a. Sample answer: The *biotic factors* in an environment include both the living things and the materials that are derived from once-living things. The *abiotic factors* include all of the other aspects of the physical environment.
   b. Sample answer: A *herbivore* is an animal that consumes mainly plants. A *carnivore* is an animal that consumes mainly other animals.
   c. Sample answer: *Producers* are living things that store energy directly from the sun through photosynthesis. *Consumers* are living things that do not store energy directly from the sun and therefore must consume the producers.
   d. Sample answer: A *predator* is an animal that captures and eats another animal, which is called its *prey*.

Correction/Completion

3. Answer: niche habitat

4. Answer: scavengers

5. Answer: decomposers

6. Answer:
   a. Chloe's example is incorrect because the elodea doesn't eat minnows. Elodea is a producer, which by definition is not a consumer.
   b. C.J.'s example is correct.
   c. Austin's example is incorrect because minnows do not eat turtles.

Short Response

7. Answer:
   a. Niche
   b. Habitat
   c. Habitat
   d. Niche

8. Answer:
   a. Parasitism (The heartworm benefits at the expense of the dog.)
   b. Mutualism (Both the rhinoceros and the egret benefit.)
   c. Commensalism (The vulture benefits without harming the lion.)

9. Answer:
   a. Lettuce, tomato, dill pickle, onions, ketchup, bread
   b. Hamburger, cheese, bacon, egg
   c. Cattle eat grasses, soybeans, and animal byproducts. Chickens and pigs eat grains, plant byproducts, and meat scraps.
   d. The sun → lettuce
      The sun → tomato (and ketchup)
      The sun → cucumber (pickle)
      The sun → onion
      The sun → wheat
      The sun → grasses/soybeans/grains → cattle (hamburger, milk for cheese)
      The sun → corn/grains/soybeans → animals → pigs (bacon)
      The sun → grains/plants → animals → chickens (egg)

10. Answer:

Mutualism	Commensalism	Parasitism
Honeybee and flower	Barnacles and crab	Tapeworm and child
Algae and coral reef	Suckerfish and shark	Mistletoe and oak tree

11. Answer:
    a. Sample answers: Frozen rock and soil; polar (freezing) wind; ice and snow; limited sunlight; low rainfall
    b. Sample answer: The polar bear has thick fur and layers of fat for insulation against the freezing temperatures of its habitat. Its coat is white so that it is camouflaged by the snow.

12. Answer:
    a. Sample answer: Some birds have a diet that consists entirely of other birds.
    b. Sample answer: The diet of deer consists only of plants, such as grass, twigs, and fruit. They do not eat other animals.
    c. Sample answer: Bears are known to eat fish as well as berries.
    d. Sample answer: Ravens and other scavenging birds are sometimes observed along highways eating the meat of animals that have been killed by cars.
    e. Sample answer: Dead plants are consumed by bacteria that are too small to be seen.

13. Answer:
    Answers will vary depending on choice of animal but should include a clear and complete description of the place where the animal lives (its habitat) and its relationships with other living and nonliving things (its niche).

14. Answer:
    Answers will vary depending on choice of plant but should include a clear and complete description of the place where the plant lives (its habitat) and its relationships with other living and nonliving things (its niche).

Graphic

15. Answer:
    a. Week 4
    b. Week 12
    c. Sample answer: As the grass increases, so will the mouse population. Then the amount of grass will decrease because more mice are feeding in the same area. As the amount of grass decreases because of over-consumption, the mouse population will decline because more mice will die of starvation than will be born. When the amount of grass increases again, so will the mouse population.

16. Answer:
    Frog embryos will develop between about 2°C and 29°C, so the *range of tolerance for temperature* is 2-29°C. The *optimum temperature* is 22°C because that is the temperature at which embryo development is greatest.

17. Answer:

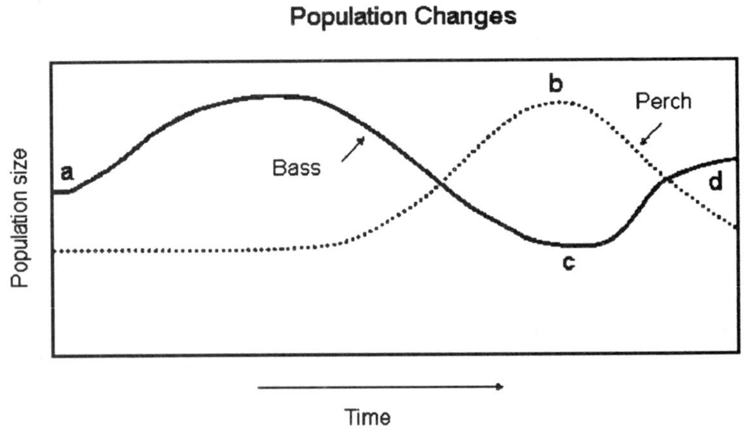

Illustrative

18. Answer:
    a. Lettuce
    b. Lettuce
    c. Every organism in the web except the lettuce
    d. Foxes, hawks
    e. Every organism in the web would die if the lettuce died because lettuce is the community's only producer. Consumers would have nothing to eat if the producer died.

19. Answer:
    a. Consumer-producer
    b. Predator-prey
    c. Consumer-producer
    d. Consumer-producer
    e. Omnivore-food
    f. Parasite-host
    g. Scavenger-food
    h. Decomposer-food

20. Answer:
a. mice Herbivores
grasses Producers
coyotes Carnivores
owls Carnivores
bacteria Decomposers
b. Sample answer: The arrows labeled *bacteria* indicate that bacteria are decomposing the owls and the mice and are returning nutrients to the soil. The grasses will use these nutrients to grow and produce their own food.

21. Answer:
a. See diagram below.
b. The grasshopper is prey for the wren and the frog. The wren and the frog are prey for the snake and are predators of the grasshopper. The snake is a predator of the wren and the frog.
c. The wren population would increase because of the greater food supply. The snake population would then increase because of the increased wren population.
d. The food web would have to include scavengers and decomposers.

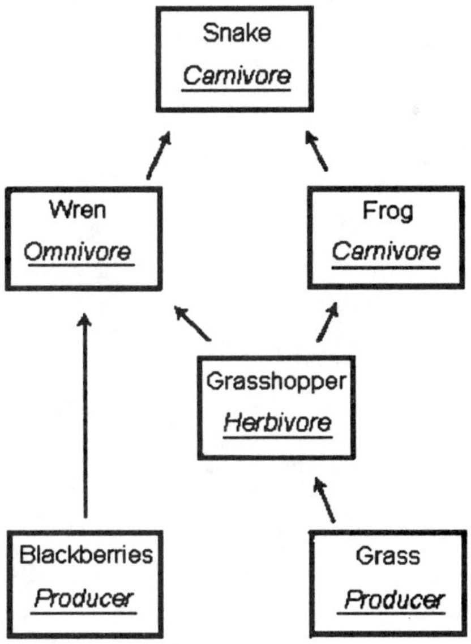

22. Answer:
Sample answers:
a. The hawk breathes air for oxygen. (abiotic-biotic)
b. The hawk eats the mouse. (biotic-biotic)
c. The mouse eats grass. (biotic-biotic)
d. The grass needs water to grow. (abiotic-biotic)
e. Moss needs water to grow. (abiotic-biotic)
f. Moss needs sunlight to grow. (abiotic-biotic)
g. Sunlight warms the air. (abiotic-abiotic)
h. The robin breathes air for oxygen. (abiotic-biotic)
i. The robin eats the earthworm. (biotic-biotic)
j. The earthworm needs water to keep its skin moist. (abiotic-biotic)

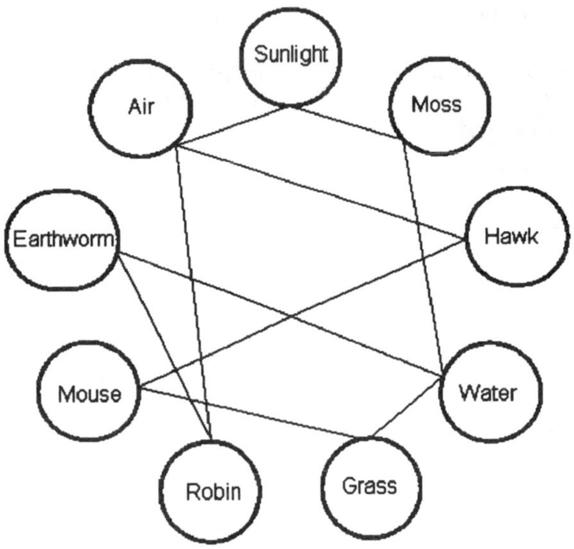

23. Answer:

D

B

A

C

24. Answer:
   Illustrations will vary but should correctly represent one of the relationships listed.
   a. Sample answer: predator-prey
   b. Sample description: A leopard is shown here attacking an antelope to represent the role of the leopard as a predator and the antelope as one of its prey.

25. Answer:
   a. See sample food web below.
   b. producers: Plankton, seaweed
      consumers: The primary consumers are herring, sand lance, and mackerel; the salmon and whale are secondary consumers.
      herbivores: Herring, sand lance
      carnivores: Salmon, whale
      omnivores: Mackerel
   c. An acceptable answer includes one of the following: The herring and the sand lance are competing for plankton; the mackerel and the herring are competing for seaweed; the salmon and the whale are competing for mackerel.
   d. An acceptable answer includes at least three of the following conclusions. An answer that includes all of the conclusions deserves a bonus. A disease that killed many mackerel would reduce the salmon and whale populations. Because there would be more seaweed for the herring, the herring population would increase. The reduction in salmon could increase the sand lance population. In turn, the plankton population could be reduced by the increased herring and sand lance populations. Increased consumption of seaweed by the herring might be balanced by a reduced number of mackerel.
   e. They would eat the remains of the dead mackerel. As salmon and whales died, the scavengers would eat their remains as well.
   f. Plankton and seaweed
   g. Sample answer: The whale is the top carnivore, consuming both mackerel and salmon. Plankton, seaweed, herring, and sand lance are other biotic factors in the environment that the whale ultimately depends on. Abiotic factors include the water and sunlight. An acceptable answer should include the relationship between the whale and two living things in its niche. An answer that mentions all of the above biotic and abiotic factors deserves a bonus.

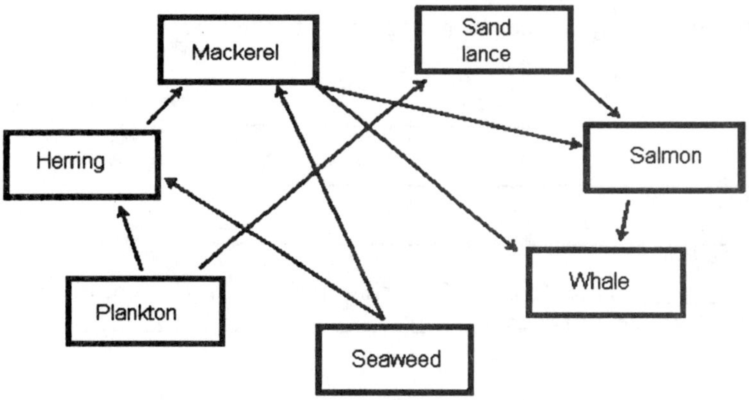

Performance Task

26. Answer:
   Answers will vary but should demonstrate an understanding of the major biotic and abiotic components of the environment for your area. Students should also effectively describe the interactions between the biotic and abiotic components of the environment and portray possible risks to the environment. Conclusions should be presented in a clear and logical report.

Chapter 4 Assessment

Word Usage

1. Use all of the following terms in one or two sentences to show how they are related:
   a. diversity, organisms, biologists, differences

   ____

   ____

   ____

   ____

   b. appearance, habitat, living things, eating, characteristics, distinguish

   ____

   ____

   ____

   ____

Correction/Completion

2. The statements below are either incorrect or incomplete. Your challenge is to make them correct and complete.
   a. The diversity of life can be seen only in certain regions of the world among specific organisms.

   ____

   ____

   ____

   b. There are more than ____________ known kinds of living things in the world, but the actual number is estimated to be between ____________ and ______________.

Short Response

3. Compare and contrast bats and birds.

Illustrative

4. Look at the illustration below. How is the diversity of life influencing these two organisms?

Short Essay

5. Suppose that a layer of dust in the atmosphere caused a worldwide reduction in the average temperature. If this condition were to last for hundreds of years, how might the diversity of life be affected in the place where you live?

Answers to Chapter 4 Assessment

Word Usage

1. Answer:
   a. Sample answer: *Biologists* use the term *diversity* to describe *differences* among *organisms*.
   b. Sample answer: Some of the *characteristics* that *distinguish living things* from one another include *habitat*, *appearance*, and *eating* habits.

Correction/Completion

2. Answer:
   a. The diversity of life can be seen *in every part of the world among all types of organisms*.
   b. There are more than <u>1.4 million</u> known kinds of living things in the world, but the actual number is estimated to be between <u>10 million</u> and <u>100 million</u>.

Short Response

3. Answer:
   Sample answer: Although bats and birds may appear to have a lot in common because they are both flying animals, they are more different than they are alike. Bats are mammals and nurse their young, while birds, which are not mammals, lay eggs. Bats tend to be nocturnal, while birds tend to be most active during the day.

Illustrative

4. Answer:
   Sample answer: The diversity of life is influencing the roles of these animals in a predator-prey relationship. The cheetah is a powerful hunter and has a body equipped for fast pursuit of its prey. The gazelle does not have the body type to compete or fight with the cheetah. These differences are essential to maintaining their roles in their environment.

   Difficulty: 3

Short Essay

5. Answer:
   Answers will vary, but students should make it clear that all aspects of the environment—living and nonliving—are tightly interconnected. When any one aspect of the environment changes, there will be repercussions such as the extinction of a species or the evolution of a new species through adaptation. It is difficult to predict with any certainty what those repercussions will be.

   Difficulty: 4

Chapter 5 Assessment

Word Usage

1. Use all of the following terms in one or two sentences to show how they are related:
   a. adaptations, organisms, natural selection, reproduce, environments

   b. camouflage, mimicry, changes, snowshoe hare, avoid, hornet fly

Correction/Completion

2. The statements below are either incorrect or incomplete. Your challenge is to make them correct and complete.
   a. Members of a species can display adaptations to a change in their environment within a short period of time.

      ____

      ____

      ____

      ____

   b. ______________ believed that an organism could acquire a favorable characteristic during its lifetime and pass that characteristic on to its offspring.
   c. ______________ believed that organisms with favorable characteristics were more likely to survive, reproduce, and pass those favorable characteristics to their offspring.

Short Response

3. For each adaptation on the table below, mark an "X" to show whether the adaptation helps primarily with locomotion, protection, or getting food.

Adaptation	Locomotion	Protection	Obtaining food
Eagles have extremely sharp vision.			
Some monkeys can hold branches with their tails.			
Roosters have a sharp spur on each foot.			
Ivy plants have flat, dark green leaves.			
Mesquite trees grow spikes on their branches.			
Bats have thin membranes between their fingers.			

Numerical Problem

4. Every spring, Nina begins to work on her lawn and gardens. First she counts the number of dandelions on her lawn. She records the number each year in order to determine how this particular area resists or succumbs to weeds. Two years ago, there were 18 dandelions, last year there were 23 dandelions, and this year there are 28 dandelions.
   a. If the dandelions keep growing at the same rate, how many dandelions will there be next year?

   ____

   ____

   b. Nina is frustrated by the greater amount of weeding she must do every year. Explain why the dandelions are so successful at reproducing.

   ____

   ____

   ____

   ____

   ____

   ____

Short Essay

5. Applying the theory of natural selection, explain how the ancestors of the air-breathing porpoise, who were land animals, developed the ability to live in the water.

Answers to Chapter 5 Assessment

Word Usage

1. Answer:
   a. Sample answer: According to the theory of *natural selection*, *organisms* with *adaptations* that are best suited to their *environments* are more likely to survive and to *reproduce* at a higher rate than other organisms.
   b. Sample answer: *Camouflage* helps the *snowshoe hare* survive in its changing environment because its coat *changes* color to match its surroundings. *Mimicry* protects the *hornet fly* from predators because it looks like a hornet, which predators *avoid* because hornets sting.

Correction/Completion

2. Answer:
   a. Members of a species may develop adaptations to a change in their environment *over a long* period of time.
   b. Jean Baptiste de Lamarck
   c. Charles Darwin

Short Response

3. Answer:

Adaptation	Locomotion	Protection	Obtaining food
Eagles have extremely sharp vision.			X
Some monkeys can hold branches with their tails.	X		
Roosters have a sharp spur on each foot.		X	
Ivy plants have flat, dark green leaves.			X
Mesquite trees grow spikes on their branches.		X	
Bats have thin membranes between their fingers.	X		

Numerical Problem

4. Answer:
   a. There will probably be 28 + 5, or 33, dandelions.
   b. Dandelions have developed a successful adaptation for seed dispersal. The seeds are located at the tips of fine, fringelike strands that detach easily and float away. They thrive in windy areas where even gentle breezes can send the seeds to new locations, creating new dandelions.

   Difficulty: 3

Short Essay

5. Answer:
   Sample answer: Over time, the environment that the ancestors of the air-breathing porpoises lived in changed, requiring the species to live near or in the water. A few ancestors had characteristics that were favorable to living in the ocean. These ancestors were selected by their environment for survival. They could successfully reproduce, and their offspring also survived to reproduce because they inherited the characteristics that allowed them to adapt successfully to their environment. Over many generations the porpoises evolved to become ocean dwellers.

   Difficulty: 4

Chapter 6 Assessment

Word Usage

1. Use all of the following terms in one or two sentences to show how they are related:
   a. animal, subgroup, Linnaeus, vertebrates, kingdom, classification, group

   ____

   ____

   ____

   ____

   b. arthropods, lobster, invertebrates, subgroup, kingdom

   ____

   ____

   ____

   ____

Correction/Completion

2. The statements below are either incorrect or incomplete. Your challenge is to make them correct and complete.
   a. *Homo sapiens* is the scientific name for ________________.
   b. According to modern biologists, mushrooms, mosses, and maples all belong to the plant kingdom.

   ____

   ____

   ____

Short Response

3. For each animal below, write either *V* for vertebrate or *I* for invertebrate.

   ___ frog　　___ shark　　___ mosquito

   ___ lobster　　___ bat　　___ snail

4. Circle the two organisms in each group that are most closely related to each other. Then tell what important characteristic they have in common.

a. wasp earthworm eel ______________________

b. seal tuna horse ______________________

c. rosebush oak fern ______________________

d. robin bat armadillo ______________________

Illustrative

5. Look at the marine environment below and then use the space that follows to make a classification diagram for the organisms shown.

Short Essay

6. Why do scientists find it useful to have a consistent system for classifying living things?

Answers to Chapter 6 Assessment

Word Usage

1. Answer:
   a. Sample answer: The largest *group* of living things in *Linnaeus's* system of *classification* is the *kingdom*. One *subgroup* of the *animal* kingdom is the *vertebrates*—animals with a backbone.
   b. Sample answer: A *lobster* is a member of the *subgroup* crustaceans, which is a subgroup of *arthropods*, which is a subgroup of *invertebrates*, which is a subgroup of the animal *kingdom*.

Correction/Completion

2. Answer:
   a. human beings
   b. According to modern biologists, mosses and maples belong to the plant kingdom, *but mushrooms belong to the other (fungi) kingdom.*

Short Response

3. Answer:

V frog	V shark	I mosquito
I lobster	V bat	I snail

4. Answer:

a.	*wasp*	*earthworm*	eel	Both are invertebrates.
b.	*seal*	tuna	*horse*	Both are mammals.
c.	*rosebush*	*oak*	fern	Both are flowering plants.
d.	robin	*bat*	*armadillo*	Both are mammals.

Illustrative

5. Answer:

Sample answer:

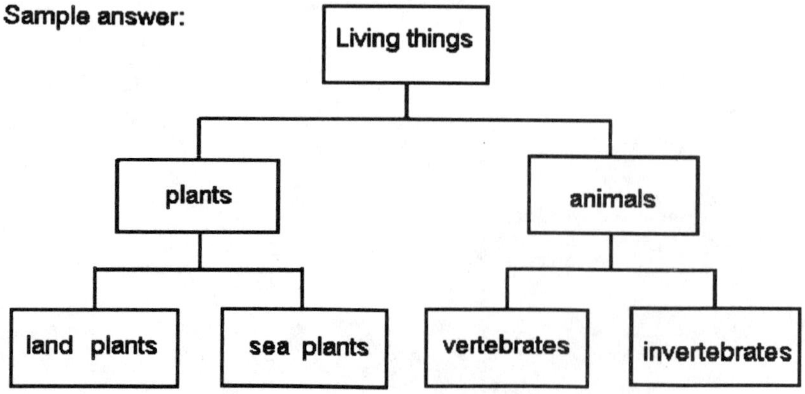

Difficulty: 3

Short Essay

6. Answer:
   Sample answer: By using a consistent classification system, scientists not only keep track of millions of different organisms but also learn more about them. Using one language to name organisms eliminates confusion among international scientists when referring to a specific organism. Since the classification system requires studying each organism's interior as well as its exterior, scientists learn how organisms carry on their life functions and how they are related to other organisms. For example, whales look much like fish, but because they have lungs and mammary glands, they are actually more closely related to humans.

   Difficulty: 4

Unit 2 End-of-Unit Assessment

Word Usage

1. Use all of the following terms in one or two sentences to show how they are related.
   a. Charles Darwin, finches, natural selection, adaptations

   b. attract, hornet, mate, sting, peacock, adaptations, predators

2. The woolly mammoth was an elephantlike animal with a heavy coat. It lived in North America about 10,000 years ago, until the end of the last ice age. Use the terms below to make a hypothesis about the disappearance of these animals.

   extinct, environmental change, natural selection, adapt

Correction/Completion

3. The statements below are either incorrect or incomplete. Your challenge is to make them correct and complete.
   a. Arthropods are successful vertebrates because they survive in many different habitats.

   ____

   ____

   ____

   b. Scientists use a ____________________ system to show the evolutionary relationship of organisms.

Short Response

4. Using an example, describe an adaptation that a plant or animal has for the following:
   a. "disappearing" into its environment

   ____

   ____

   ____

   ____

   ____

   ____

   ____

   b. storing water

   ____

   ____

   ____

   ____

   ____

   ____

   ____

5. Seeds can be transported by wind, water, animals, or mechanical propulsion (like a bullet). Describe an adaptation that a seed might have for transportation by one of these methods.

____

____

____

____

____

____

____

____

6. How might Darwin have used the following observations to support his theory of natural selection?
   a. Finches that live in different parts of the Galápagos Islands have different kinds of beaks.

____

____

____

____

____

____

____

____

   b. Fossilized seashells are found high in the Andes Mountains of South America.

____

____

____

____

____

____

____

Data for Interpretation

7. In the table below are some characteristics for students in an eighth-grade class. Use the table to answer the questions that follow.

Characteristic		Van	Tasha	Tiffany	Bo	Josh
Hand folding, thumb position	left over right		X		X	X
	right over left	X		X		
Ear lobes	attached	X	X			X
	free			X	X	
Hairline	pointed		X		X	X
	straight	X		X		
Tongue rolling	can roll	X	X			
	can't roll			X	X	X

a. How many have three identical characteristics? Who are they?

____

____

b. How many have only two identical characteristics? Who are they?

____

____

c. How many have only one identical characteristic? Who are they?

____

____

d. Are there any two people who are different in every respect? If so, who are they?

____

____

e. How does this data relate to diversity?

____

____

____

Graphic

8. The graph estimates the number of named animal species in the world. Use the graph to answer the questions.

   a. What is the total number of different kinds of animals?

   ____

   ____

   b. Which animals are in the majority?

   ____

   ____

   c. There are about ____________ times as many arthropods in the world as all other animals put together.

   d. What can you conclude about the diversity of the animals in this graph?

   ____

   ____

   ____

   ____

   ____

   ____

   ____

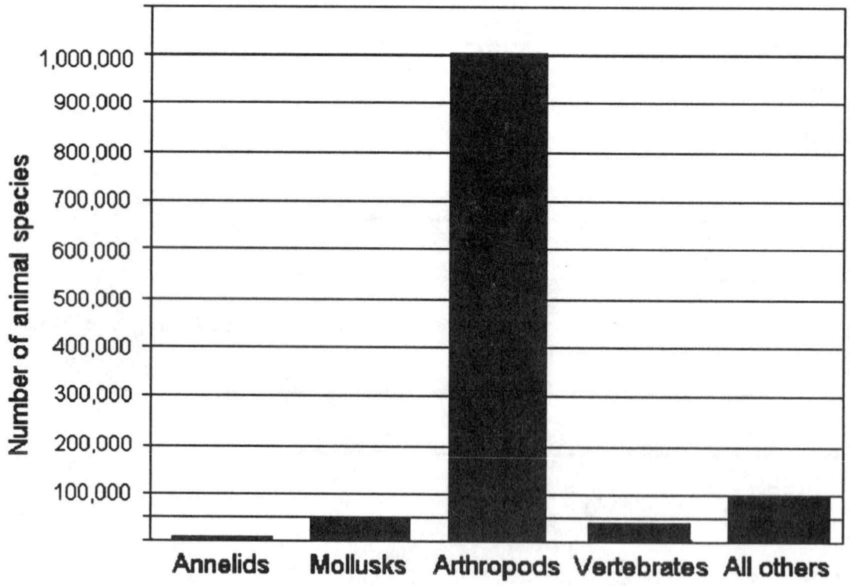

Numerical Problem

9. Suppose that the diversity of life on Earth were measured by percentages and that 1.4 million species corresponds to 100 percent.
   a. What does this mean?

   b. There are 250,000 species of angiosperms, or flowering plants. What percent of the total existing species on Earth do angiosperms make up?

   c. How would the destruction of all angiosperms affect the diversity of life on Earth?

Illustrative

10. Describe how each of these animals is adapted to survive in its environment. Mention adaptations for obtaining food, for protection, or for locomotion.

a. ____

____

____

____

____

b. ____

____

____

____

____

c. ____

____

____

____

____

a. b. c.

Short Essay

11. What are some of the ways in which you benefit from the diversity of life?

____

____

____

____

____

____

____

____

____

Illustrative

12. Examine the living things pictured, and answer the questions.
    a. Create a chart to classify these organisms into groups and subgroups. Then divide each of the subgroups so that each organism has its own subgroup. Explain the characteristic that distinguishes each organism in your classification system.
    b. According to classification systems used by biologists today, why is a butterfly more closely related to an earthworm than to an ostrich?

____

____

____

____

Sunflower

Butterfly

Cow

Rosebush

Earthworm

Pine tree with cones

Ostrich

Fern

Answers to Unit 2 End-of-Unit Assessment

Word Usage

1. Answer:
   a. Sample answer: *Charles Darwin* studied *adaptations* in *finches* when forming his theory of *natural selection*.
   b. Sample answer: Some animals have *adaptations* that help them *attract* attention. A male *peacock* displays his feathers to tempt his *mate*, while the black-and-yellow pattern of a *hornet* warns enemies and *predators* of its powerful *sting*.

2. Answer:
   Sample hypothesis: After the last ice age, there were great *environmental changes* as the climate became warmer and the icecaps melted. According to the theory of *natural selection*, those species that could not *adapt* to the warmer climate, like the woolly mammoth, were not selected for survival and became *extinct*.

Correction/Completion

3. Answer:
   a. Arthropods are successful *invertebrates* because they survive in many different habitats.
   b. classification

Short Response

4. Answer:
   a. Sample answer: Camouflage enables an organism to "disappear" into its surroundings. Examples will vary. The walking stick is an insect that looks so much like a stick that it often goes undetected by predators.
   b. Sample answer: Specialized parts of plant cells enable plants to store a lot of water. Cactuses store water in thickened, fleshy stems that have reduced or spinelike leaves and a heavy, waxy coating that prevents water loss.

5. Answer:
   Sample answers: Seeds that have wings or tufts could be carried by wind. Seeds that have waterproof air sacs could float on and be transported by water. Seeds that have a tough coat that resists digestion could be dispersed by means of animal feces. Seeds positioned on a high, flexible stalk could be thrown from the plant and transported by mechanical propulsion.

6. Answer:
   a. Finches living in different parts of the Galápagos Islands have similar features of a common ancestor but have evolved different types of beaks to eat the food available in different areas of their environment. Only those individuals with beak adaptations that allowed them to eat the available food survived to reproduce.
   b. The fact that fossilized seashells are found in the Andes Mountains shows how drastically an environment can change. An area that once was populated by ocean-living organisms changed over time to become mountainous terrain. Organisms must adapt to a changing environment in order to survive.

Data for Interpretation

7. Answer:
   a. Two pairs of students: Tasha and Josh, and Bo and Josh
   b. Four pairs of students: Van and Tasha, Van and Tiffany, Tasha and Bo, and Tiffany and Bo
   c. Two pairs of students: Van and Josh, and Tiffany and Josh
   d. Two pairs of students: Van and Bo, and Tasha and Tiffany
   e. Sample answer: The diversity of living things is due to the many different combinations of characteristics within a species.

Graphic

8. Answer:
   a. There are almost 1.2 million named species of animals in the world.
   b. Arthropods are in the majority, with almost 1 million different species.
   c. five
   d. Sample answer: The number of different animals in each group shows the diversity of the group. Arthropods are the most diverse because there are so many species. These organisms have the ability to adapt to their environments and reproduce successfully. The other animals shown in the graph do not have the level of diversity that the arthropods do.

Numerical Problem

9. Answer:
   a. Having a diversity of 100 percent with 1.4 million organisms suggests that all habitats and niches are fulfilled by the existing organisms.
   b. $\frac{(250{,}000)}{(1{,}400{,}000)} = \frac{(17.9)}{(100)}$ Angiosperms make up about 17.9 percent of the existing species on Earth.
   c. The diversity measurement would change from 100 percent to 100 - 17.9, or 82.1, percent. It is possible that the loss of angiosperms would mean the loss of other, dependent, species. According to the theory of natural selection, however, all habitats and niches made available by these losses would eventually be filled by other organisms, and the diversity measurement would once again be 100 percent.

Illustrative

10. Answer:
   a. Since the rabbit is a herbivore, it must have adaptations for safe grazing or for grazing without being attacked by predators. It has peripheral vision and long ears that enable it to see and hear approaching predators. It also has strong leg muscles that allow it to move quickly to escape an attack by a predator.
   b. The chimpanzee eats mainly fruits, leaves, and seeds and uses its hands to hold tools to search for termites and ants. The long arms of a chimpanzee enable it to climb from tree to tree to escape predators as well as to nest in the branches.
   c. The mountain lion has large paws, sharp teeth, and speed to capture its prey. It runs quickly because it has strong leg muscles. The mountain lion has no natural predators.

   Difficulty: 3

Short Essay

11. Answer:
   Answers will vary, but students should demonstrate an understanding of the amount of diversity even in their local environments. They may wish to focus on the interdependence of organisms in terms of food acquisition, or they may address the balance that exists between various organisms in nature.

   Difficulty: 4

Illustrative

12. Answer:
   a. See sample classification chart below.
   b. Sample answer: A butterfly is more closely related to an earthworm than to an ostrich because both the butterfly and the earthworm are invertebrates, while the ostrich is a vertebrate.

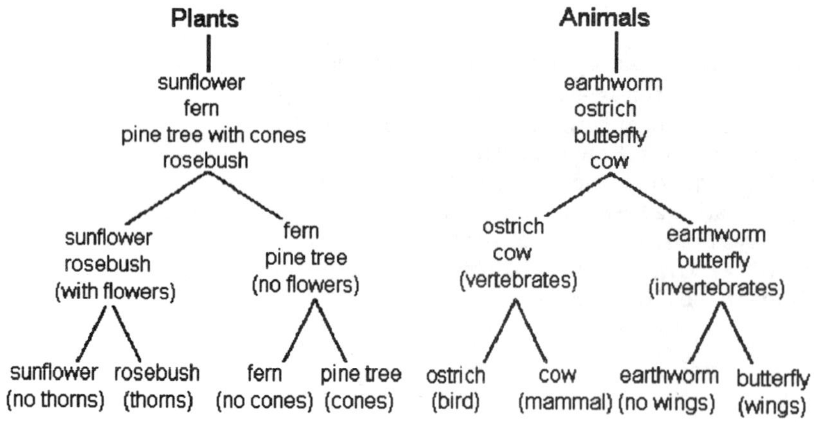

   Difficulty: 4

Activity Assessment

1. **Exploring Diversity**

Teacher's Notes

Overview
Students analyze different kinds of leaves and classify them by their similarities. They then create a poster detailing their classification system.

Materials
(per activity station)
- 8–10 different kinds of plant leaves (consider using 3 leaves from trees, 3 from shrubs or bushes, and 3 from fruits or vegetables)
- scissors
- colored pencils
- wax paper
- a metric ruler
- transparent tape
- a magnifying glass
- poster board

Preparation
Prior to the assessment, equip student activity stations with the materials needed for each experiment. Remind students not to taste the plant leaves.

Time Required
Each student should have 15 minutes at the activity station and 30 minutes to complete the Data Chart and poster.

Performance
At the end of the assessment, students should turn in the following:
- a completed Data Chart
- a poster detailing a classification system

Evaluation
The following is a recommended breakdown for evaluation of this Activity Assessment:
- 10% use of equipment and preparation of leaves
- 40% quality and clarity of observations
- 40% design of a clear classification system
- 10% neatness and clarity of poster

Safety Alert!

2. **Exploring Diversity**

Imagine that you have been transported back in time to the eighteenth century. You have been selected to work with scientist Carolus Linnaeus as he develops the first classification system for plants and animals. You've been asked to examine several plant leaves and to decide which factors are important in classifying them. What recommendations will you make to Linnaeus?

Before You Begin . . .

As you work through the tasks, keep in mind that your teacher will be observing:

- how you prepare the leaves for examination
- your ability to record your observations clearly
- how well you classify the leaves by their similarities
- how clearly you present your classification system on the poster

Now you are ready to begin your eighteenth-century adventure!

Task 1 Cut the wax paper into sheets. Press one leaf between two sheets of wax paper and tape the edges of the papers together. Repeat this procedure for all of the leaves at your workstation. Then assign each leaf a number so that you can distinguish each of your samples.

Task 2 Examine each leaf. What are its basic characteristics? Record your observations on your Data Chart.

Task 3 Analyze your Data Chart for similarities among the leaves. Classify the samples by placing similar leaves in the same group.

Task 4 Outline your classification system on a piece of poster board. Tape similar samples next to each other with a brief explanation of what the leaves have in common and how they are different from the rest of the leaves.

Safety Alert!

3. **Classification System Data Chart**

Leaf number	Color	Size	Shape	Other features

Answers to Unit 2 Activity Assessment

1. Answer: Not applicable (teacher's notes)

2. Answer: Not applicable (student's notes)

3. Answer:
   Answers will vary depending on samples used, but students should demonstrate the ability to adequately prepare the leaves, make thorough observations about the leaves, and classify the leaves based on similarities. The classification system should be clearly and accurately presented on a poster board.

Unit 2 SourceBook Assessment

Short Response

1. Fill in the blanks with the following subcategories, in order, beginning with kingdom and ending with the most specific classification: class, family, species, phylum.

   kingdom, ________, ______, order, _______, genus, ______

2. Put the following terms in the proper order, from the longest period of time to the shortest: era, decade, period, week, year.

Multiple Choice

3. Monerans, protists, animals, and fungi are examples of the category
   a. phylum. b. class. c. species. d. kingdom.

4. Cyanobacteria contain chlorophyll and produce their own food by a process called
   a. respiration. b. photosynthesis. c. conduction. d. osmosis.

5. Protozoans are animal-like protists; therefore, they (do/do not) have a cell wall, and they (do/do not) contain chlorophyll. (Circle the correct words.)

6. The black or gray powdery substance that you may find growing on an old hamburger bun is an example of a(n)
   a. fungus. b. cyanobacterium. c. alga. d. moneran.

7. Redwood trees are
   a. vascular plants. b. nonvascular plants.

8. The most diverse group of modern plants are those with
   a. "naked" seeds. b. "covered" seeds.

9. Which of the following does NOT correctly complete the sentence?
   You are
   a. an invertebrate. b. an animal. c. warmblooded. d. a mammal.

10. Animals must
    a. be multicellular. b. have organ systems. c. have vertebrae.
    d. be able to make their own food within their body.
    e. All of the above f. a and b only g. a, b, and d only

11. A man walked into a restaurant and ordered an arthropod. Which of the following did he order?
    a. steak b. catfish c. lobster d. chicken

12. A person's bones are called an
    a. endoskeleton. b. exoskeleton.

13. Coldblooded refers to
    a. animals having a body temperature that changes according to the temperature of the environment.
    b. all vertebrates that live in the water.
    c. animals having a body temperature below 20°C.
    d. a and b only

14. Which invertebrates have gills when they are young and lungs when they mature? These animals also live part of their lives in water and part on land.
    a. reptiles
    b. coelenterates
    c. amphibians
    d. segmented worms

15. The term *millennium* means
    a. 10,000 years.
    b. 5 decades.
    c. any long period of time.
    d. 1000 years.

16. A blue jay laid two eggs. The first egg hatched a very strong and healthy female blue jay. The second egg hatched a much weaker and smaller female blue jay. The first blue jay was able to learn to fly, mate, and reproduce. The second was unable to fly very well and died in just a few weeks. What process is described here?
    a. fate b. natural selection c. adaptation d. mutation

17. Farmer Jane grew tomatoes. She took the seeds from the largest, reddest, and best-tasting tomatoes and planted them. From the plants that grew, she again took the seeds from the largest, reddest, and best-tasting tomatoes and planted them. She repeated this for three seasons. Farmer Jane then had the largest, reddest, and best-tasting tomatoes in the whole county. This scientific process is known as
    a. survival of the fittest. b. resistance.
    c. selective breeding. d. variation.

Short Response

18. Fishermen once tried to eliminate starfish by cutting them into pieces and throwing them back into the sea. This, however, did not solve their problem. It only made it worse. What characteristic of starfish (an echinoderm) explains why the situation worsened?

____

____

____

____

19. If you were digging deep in your backyard and found a piece of stone with the imprint of a starfish, could you find a scientific explanation for how it got there? (Your neighbor burying it there is not a scientific explanation.)

____

____

____

____

____

____

____

____

20. Why are albino Bengal tigers more scarce than non-albino Bengal tigers? (Assume that albinism is just as common as non-albinism in Bengal tigers.)

____

____

____

____

____

____

21. The bones of a person's arm and the bones of a whale's flipper are arranged similarly, but they differ in size and shape. What can you infer from this observation?

____

____

____

____

True/False

22. Only mammals are warmblooded.
    a. true b. false

23. Birds have hollow bones that better enable them to fly.
    a. true b. false

24. From what we have learned about the history of the Earth, mammals were among the first organisms to appear on Earth.
    a. true   b. false

25. All organisms in the kingdom Protista are multicelled.
    a. true   b. false

Answers to Unit 2 SourceBook Assessment

Short Response

1. Answer: kingdom, phylum, class, order, family, genus, species

2. Answer: era, period, decade, year, week

Multiple Choice

3. Answer: d. kingdom.

4. Answer: b. photosynthesis.

5. Answer:
   Protozoans are animal-like protists; therefore, they (do/*do not*) have a cell wall, and they (do/*do not*) contain chlorophyll. (Circle the correct words.)

6. Answer: a. fungus.

7. Answer: a. vascular plants.

8. Answer: b. "covered" seeds.

9. Answer: a. an invertebrate.

10. Answer: f. a and b only

11. Answer: c. lobster

12. Answer: a. endoskeleton.

13. Answer:
    a. animals having a body temperature that changes according to the temperature of the environment.

14. Answer: c. amphibians

15. Answer: d. 1000 years.

16. Answer: b. natural selection

17. Answer: c. selective breeding.

Short Response

18. Answer:
Regeneration; each piece underwent regeneration and became a complete individual, which greatly increased the number of starfish.

19. Answer:
Answers may vary, but should involve a description of the Earth's changes over time. For example, one answer may be that the backyard was covered by ocean at one time.

20. Answer:
Answers may vary, but the following idea should be included: The white color makes albino Bengal tigers less able to camouflage themselves in their environment; therefore, they are more apt to become prey (of hunters) or, on the other hand, to be less successful in stalking their prey.

21. Answer: Students may infer that the whale and the human may have had a common ancestor.

True/False

22. Answer: b. false

23. Answer: a. true

24. Answer: b. false

25. Answer: b. false

Unit 2 Extra Assessment Items

Word Usage

1. For each group of words below, write one or two sentences to show how the words are related to one another.
   a. mimicry, adaptations, survive, predators

   b. scientific name, *Homo sapiens*, Linnaeus, you

   c. group, subgroup, kingdom, vertebrates, mosses, animals, plants

Correction/Completion

2. Decide whether the following statement is true or false. If it is false, rewrite the statement so that it is correct.
   Natural selection is a quick process that helps an organism survive.

Short Essay

3. Explain Darwin's theory of natural selection in such a way that younger students could understand it. In your explanation, include some of the evidence that supports this theory. Use the example of polar bears, whose ancestors may have been brown, or dolphins, whose ancestors probably lived on land.

4. Imagine that global warming caused a world-wide increase in the average temperature of the Earth. Suppose that this increase caused the following to happen in your region:
   - There are no longer freezing temperatures at any time of the year.
   - The hottest days are no hotter than at present.
   - This new situation is expected to last for hundreds of years.
   - Your region remains above sea level.

   Now predict some of the changes that could occur in the plant and animal life in your region. Describe plants and animals that would probably increase in number and some of those that would probably decrease in number, and explain why.

5. Describe how humans might adapt to global warming. Assume that global warming results in rising sea levels that cause many of the world's populated regions to flood. Mention some of the adaptations that would be likely to occur in humans.

6. In what group of animals would you classify the following creature? Explain your reasoning.

A kulse has a short, hairy body. The kulse has four jointed legs and one pair of eyes. It has a spinal cord that runs down its back. It has an internal skeleton for better movement and lives in the desert. Its brain is well developed, and it is intelligent. Female kulses are able to produce milk.

Short Response

7. Using one example, describe an adaptation that a plant or animal has for the following:
   a. breathing in water

   b. avoiding a predator

8. Use an example to show that you understand the meaning of the term *adaptation*.

9. In comparison with all other animals, what advantages do humans have that enable us to adapt to a wide variety of environments?

10. There are numerous examples of structures and behaviors that are adaptations for protection. Complete the table below by naming a protective adaptation for each living thing and by adding two other examples.

Living thing	Adaptation
skunk	
rosebush	
lion	
snake	

11. Many different kinds of adaptations for locomotion exist within the animal kingdom. Complete the table by naming the adaptations for locomotion of the animals listed and of two additional animals of your own choosing.

Living thing	Adaptation
trout	
deer	
robin	
duck	

12. Provide an example of an organism that can do the following:
   a. blend into its environment

   ____

   b. crush hard food

   ____

   c. fly from one flower to the next

   ____

Illustrative

13. Describe how the pine tree shown here is adapted to survive in its environment. Include adaptations that enable the plant to obtain light, get water and nutrients, withstand the wind, and reproduce.

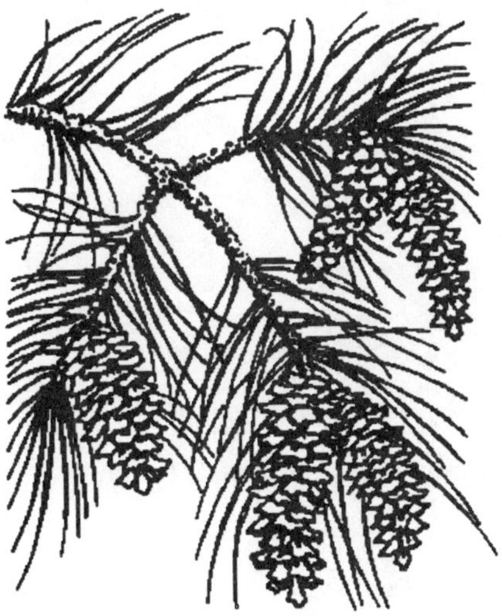

14. Describe how the dandelion shown here is adapted to survive in its environment. Include adaptations that enable the plant to obtain light, get water and nutrients, withstand the wind, and reproduce.

15. Sketch a feature or features of a seed or plant that would enable the seed to be carried by the wind.

16. Sketch a feature or features of a plant that would help it to withstand high winds.

17. Sketch a feature or features of a small animal that would help it avoid being detected or caught by a predator.

Performance Task

18. Use the classification system for the animal kingdom that you learned in this unit to classify the animals that you find at a zoo or animal park in your area.

19. To help younger students learn more about the diversity of plants in their local environment, create a nature walk. Choose an existing public trail or path, create a map of the path, and identify on the map the location of a variety of plants. Include representative examples of the following subgroups if possible: mosses, ferns and relatives, cone-bearing plants, and flower-bearing plants.

20. As your contribution to a study of the flora (plants) and fauna (animals) in your environment, choose a vacant lot, park, beach, or other location for study. Select 10 of the most commonly occurring living things in that location, including at least three plants and three animals. In your report, sketch each of the living things and describe in writing how it is adapted to its environment.

21. Are you curious about the plant and animal life in other parts of the United States or Canada? If so, you probably can exchange information with a *SciencePlus* student from the state, province, or region you are interested in. Here is what you should do.
    a. Ask your regional representative of Holt, Rinehart and Winston, Inc. (the publisher of *SciencePlus*) for the name and address of a school using *SciencePlus* in the region you are interested in. Write to the principal of that school to request a pen pal.
    b. You and your pen pal should each make sketches or collect photographs and illustrations of 10 of the most interesting plants and animals in your own area.
    c. Write something about each of these organisms based on your personal experience.
    d. Exchange this information with your pen pal.

Answers to Unit 2 Extra Assessment Items

Word Usage

1. Answer:
   a. Sample answer: *Adaptations* are features of an organism that help it *survive* in its environment. *Mimicry* is one kind of adaptation that helps an organism avoid *predators*.
   b. Sample answer: In the classification system that *Linnaeus* devised, the *scientific name* for *you* and all other humans is *Homo sapiens*.
   c. Sample answer: The largest *group* of living things in Linnaeus's system of classification is the *kingdom*. *Plants* and *animals* are classified in separate kingdoms. One *subgroup* of the animal kingdom is the *vertebrates*, while one subgroup of the plant kingdom is the *mosses*.

Correction/Completion

2. Answer:
   False. Sample answer: Natural selection is a *slow* process *in which the species whose members have adaptations that best fit their environment are the species that survive.*

Short Essay

3. Answer:
   Acceptable answers should describe a change that has occurred in a species, such as the change in fur color of polar bears or the ability of dolphins to live in water, and that has given them some advantage for survival. Give the answer a bonus if it mentions mutations as a likely cause of the initial change in a member of a species and conveys the understanding that the process of selection takes place over a very long time, much longer than the lifetime of one member of the species.

4. Answer:
   Answers will vary depending on the region in which you live. In general, though, plants and animals that are found in a warmer climate will enter your region or increase in number. Those that depend on a colder climate may disappear from your area or be reduced in number. Student explanations of why those plants and animals would increase or decrease in number should mention the specific adaptations of these plants and animals to climatic conditions. An acceptable answer should include at least one example of a plant or animal and should give a reason for the predicted effect. If one example of each is given and if competition with other plants and animals is given as one of the reasons for the prediction, give the answer a bonus.

5. Answer:
   An acceptable answer includes at least one adaptation that would likely occur in humans. If three or more adaptations are described, give the answer a bonus.

6. Answer:
Because it has locomotion, the kulse should be placed in the animal kingdom. Because it has a skeleton, it should be classified within the vertebrate subgroup of the animal kingdom. Because it has hair on its body and the female kulse is able to produce milk, this vertebrate should be classified as a mammal. An answer that includes the kulse among the vertebrates or among the mammals and that gives at least one distinguishing characteristic is acceptable. If the answer mentions that the kulse should be classified as a mammal because of its hair and the ability of the female to produce milk, it deserves a bonus.

Short Response

7. Answer:
   a. Sample answer: Gills enable animals to breathe in water. Most fish display this adaptation.
   b. Sample answer: Mimicry and camouflage enable many organisms to avoid predators. The stingless syrphus fly protects itself from predators by mimicking the yellow jacket, a wasp that predators avoid because it stings its foes.

8. Answer:
An acceptable answer provides an example of a feature that helps an organism to survive in its environment.

9. Answer:
Acceptable answers refer to human inventiveness, intelligence, tool-making ability, or examples of these. Give a bonus if two or more reasons or examples are given.

10. Answer:

Living thing	Adaptation
skunk	Sample answer: Skunks protect themselves by spraying a potent liquid when threatened.
rosebush	Sample answer: Rosebushes have thorns to protect themselves from predators.
lion	Sample answer: The large size and great strength of the lion protect it from possible foes.
snake	Sample answer: Some snakes protect themselves from threatening organisms by injecting them with a poisonous liquid.

11. Answer:

Living thing	Adaptation
trout	Sample answer: Trout are adapted to swim in streams.
deer	Sample answer: Deer are known for their ability to use speed to evade predators.
robin	Sample answer: Robins are able to fly, hop, and walk.
duck	Sample answer: Ducks are able to swim and walk on the ground.

12. Answer:
    a. Sample answer: tiger
    b. Sample answer: wolf
    c. Sample answer: bee

Illustrative

13. Answer:
An answer that describes at least one way that the plant is adapted to survive in its environment is acceptable. Any answer that describes three or more ways that the plant is adapted to survive in its environment deserves a bonus.

14. Answer:
An answer that describes at least one way the plant is adapted to survive in its environment is acceptable. Any answer that describes three or more ways that the plant is adapted to survive in its environment deserves a bonus.

15. Answer: Illustrations could include a variety of examples, such as a maple or dandelion see

16. Answer:
Illustrations could include a variety of examples, such as the root system of a plant; the flexibility or strength of a plant's stem or trunk; or a low, streamlined form.

17. Answer:
Illustrations could include a large variety of examples of camouflage, including colors that blend with the environment and designs that make the animal look like other animals, such as the "eyes" on some butterfly wings.

Performance Task

18. Answer:
Student methods of classification should be clear and concise and should enumerate as many subgroups as possible.

19. Answer: Maps should clearly and accurately reflect examples of as many subgroups as possibl

20. Answer:
Sketches should be clear and concise, and written descriptions should accurately identify the adaptations of each organism to its environment.

21. Answer: Encourage students to share their experiences with the class.

Chapter 7 Assessment

Word Usage

1. Use all of the following terms in one or two sentences to show how they are related.
   a. between, dissolves, particles, liquid, spaces, solid, evenly

   b. alloy, copper, solution, brass, solid, zinc

Correction/Completion

2. The statements below are either incorrect or incomplete. Your challenge is to make them correct and complete.
   a. The Tyndall effect can be observed when the path of a bright light is visible as it shines through a liquid that is a solution.

   b. When a spoonful of sugar is dissolved in a cup of water, the sugar is the solvent, and the water is the solute.

Short Response

3. In each sentence below, identify the solvent.
   a. Rubbing alcohol contains 30 percent water and 70 percent alcohol.

   ____

   b. Scented candles are made with a few drops of an aromatic oil and wax.

   ____

   c. Dental fillings are an alloy of silver with some mercury added to it.

   ____

   d. Stainless steel is an alloy that includes 77 percent iron, 15 percent chromium, 8 percent nickel, and less than 1 percent carbon.

   ____

Data for Interpretation

4. Val and Pedro conducted an experiment to see what effect temperature has on how well sugar dissolves in water. They recorded the different conditions for each trial in the table below.

Trial	Amount of sugar	Amount of water	Temp. of water	Amount of time	Number of stirs
1	200 g	500 mL	75°C	60 sec.	10
2	1000 g	250 mL	80°C	20 sec.	5

After they conducted their experiment, they found that no undissolved sugar remained in Trial 1 and that some undissolved sugar remained in Trial 2. Answer the following questions based on the table and the results.

a. What might they *incorrectly* conclude from these results?

____

____

____

____

____

____

b. Which variables should they have controlled?

____

____

____

____

____

c. What other suggestions would help Val and Pedro improve their experiment?

____

____

____

____

____

____

Short Essay

5. Water is a good solvent for most things, but not for everything. Some types of paint can be washed away with water before they dry, but other types of paint require a different solvent, such as turpentine. Why do you think this is so?

Answers to Chapter 7 Assessment

Word Usage

1. Answer:
   a. Sample answer: When a *solid dissolves* in a *liquid*, the *particles* of the solid spread out *evenly* and occupy *spaces between* the liquid particles.
   b. Sample answer: *Brass* is an *alloy* made from a *solid solution* of *copper* and *zinc.*

Correction/Completion

2. Answer:
   a. The Tyndall effect can be observed when the path of a bright light is visible as it shines through a liquid that is *not* a solution.
   b. When a spoonful of sugar is dissolved in a cup of water, the sugar is the *solute*, and the water is the *solvent*.

Short Response

3. Answer:
   a. 70 percent alcohol
   b. Wax
   c. Silver
   d. 77 percent iron

Data for Interpretation

4. Answer:
   a. They might incorrectly conclude that sugar dissolves better in cooler water than in warmer water because there was no undissolved sugar in Trial 1.
   b. They should have kept the amount of sugar, the amount of water, the amount of time, and the number of stirs the same for each trial. That way the only variable is temperature, and they could accurately determine its effect on the solubility of sugar.
   c. Sample answer: They could improve their experiment by creating a significant difference in the water temperature (greater than 5°C). They should also repeat their experiment several times.

   Difficulty: 3

Short Essay

5. Answer:
   Sample answer: Some paints are water-based. They will wash away in water until they are dry. Other paints are oil-based. Since water cannot dissolve oil, it cannot dissolve oil-based paint. Solvents such as turpentine and alcohol can dissolve oil, and therefore they can dissolve oil-based paints.

   Difficulty: 4

Chapter 8 Assessment

Word Usage

1. Use all of the following terms in one or two sentences to show how they are related:
   a. salt, water, desalination, solvent, solute

   ____

   ____

   ____

   b. different, chromatography, solution, ink, color

   ____

   ____

   ____

Correction/Completion

2. The statements below are either incorrect or incomplete. Your challenge is to make them correct and complete.
   a. The boiling point of water at sea level is 100°C. At higher elevations, the boiling point is higher.

   ____

   ____

   ____

   b. Complete the following statements:

   •In the process of ___________, a liquid rapidly changes to a gas.

   •Nuclear-powered ___________ plants may someday produce our fresh water and electricity.

   •Maple sap is a _____________ of water and sugar.

   •Both scientists and detectives form _____________ to explain why certain things happen the way they do.

Graphic

3. Before a distillation experiment is begun, the solution in a flask contains 5 percent salt and 95 percent water. In the graph below, draw and label curves to show how the percentages change during the experiment. Use a dotted line for the water curve and a solid line for the salt curve.

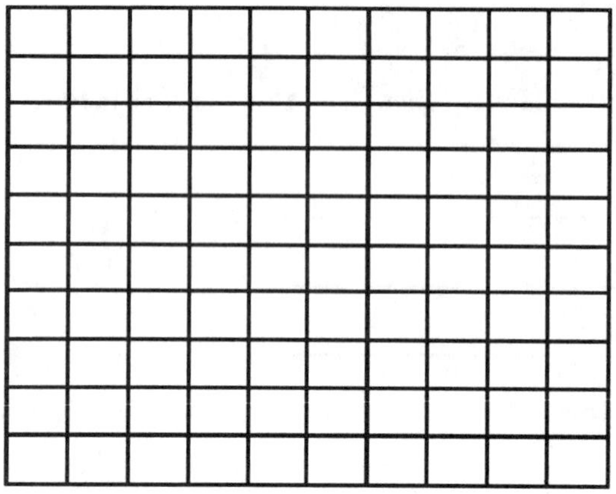

Illustrative

4. Compare illustrations *A* and *B*. Then describe three ways in which the distillation process and the water cycle are *different*.

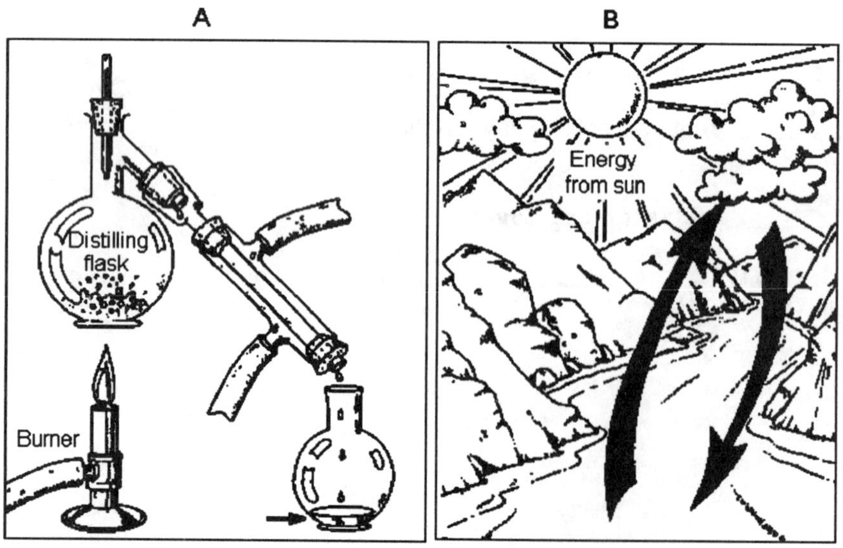

Short Essay

5. Kyle was heating water to make spaghetti. When the water boiled, he added a spoonful of salt, and he noticed that the water stopped boiling for a short time and then began to boil again. Why did a spoonful of salt interrupt the boiling, and why did it begin again?

Answers to Chapter 8 Assessment

Word Usage

1. Answer:
   a. Sample answer: During the process of *desalination*, the *solvent* (*water*) is removed and saved, and the *solute* (*salt*) is discarded.
   b. Sample answer: Paper *chromatography* is a method that can be used to find out if an *ink* is one *color* or a *solution* of *different* colors.

Correction/Completion

2. Answer:
   a. The boiling point of water at sea level is 100°C. At higher elevations, the boiling point is *lower*.
   b. •boiling
      •desalination
      •solution
      •hypotheses

Graphic

3. Answer:

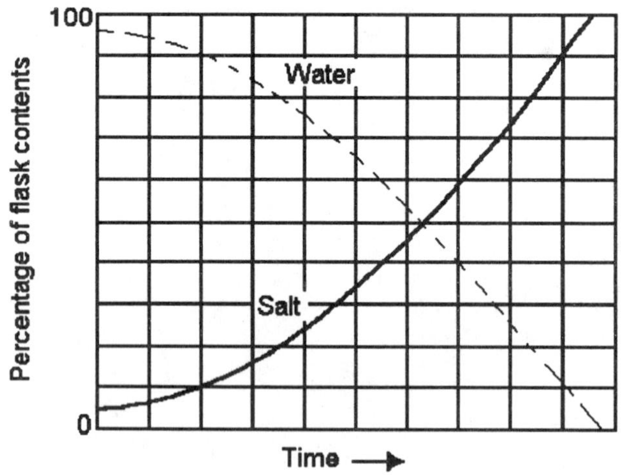

Illustrative

4. Answer:
   Sample answers: The heat source for the distillation process is a burner, but the heat source for the water cycle is the sun. In distillation the water boils, but in the water cycle it evaporates. The distillation process will stop when all of the water has boiled out of the flask, but the water cycle will continue indefinitely.

   Difficulty: 3

Short Essay

5. Answer:
   Sample answer: Water boils at 100°C at sea level. However, when a solute such as salt is dissolved in water, the boiling point increases. Therefore, the water stopped boiling when the salt was added. As the stove continued to add heat to the pot, the temperature increased until the new boiling point was reached, and the water boiled again.

   Difficulty: 4

Chapter 9 Assessment

Word Usage

1. Use all of the following terms in one or two sentences to show how they are related:
   a. solution, liquid, atmosphere, gas, ocean

   b. saturating, sugar, temperature, solubility, solution, solute

Correction/Completion

2. The statements below are either incorrect or incomplete. Your challenge is to make them correct and complete.
   a. Acid rain occurs when carbon dioxide dissolves in rainwater, which falls through the atmosphere.

      ____

      ____

      ____

   b. Concentration is the measure of the amount of solute needed to make a solution saturated.

      ____

      ____

      ____

   c. When maple syrup is made from maple sap, a solution is made more dilute.

      ____

      ____

      ____

Short Response

3. For each pair of solutions below, label one *dilute* and the other *concentrated.*

   a. __________ water from the Dead Sea

      __________ water from the ocean

   b. __________ nectar

      __________ honey

   c. __________ maple sap

      __________ maple butter

   d. __________ carbon dioxide in car exhaust

      __________ carbon dioxide in the atmosphere

Numerical Problem

4. Sea water generally contains about 3.5 g of salt and other solids for every 100 g of water. Paula is doing a biology experiment. She wants to create a model of the ocean in an aquarium. If the aquarium tank holds 40 L of water, how much salt should Paula add to create simulated sea water? Show your work.

____

____

Data for Interpretation

5. Friedrich wants to bake a cake. His recipe calls for a solution of one teaspoon of baking soda dissolved in 100 mL of water to be added to the rest of the cake mixture. Use the data below to answer the question that follows.
   - Friedrich uses tap water that is about 20°C.
   - There are 5 mL in a teaspoon.
   - There are 2.2 g of baking soda per milliliter.
   - One gram of water is equivalent to one milliliter of water.

Solubility of Baking Soda per 100 g of Water

Temperature (°C)	Baking soda (g)
0	6.9
10	8.2
20	9.6
30	11.1
40	12.7
50	14.5
60	16.4

Will Friedrich be able to make his cake under the given conditions? Explain.

____

____

____

____

____

____

____

____

Short Essay

6. Putting salt on roads will *melt* ice, but putting salt in an ice cream maker will help *freeze* ice cream. How can you explain this?

Answers to Chapter 9 Assessment

Word Usage

1. Answer:
   a. Sample answer: The *ocean* is a *liquid solution*, and the *atmosphere* is a *gas* solution.
   b. Sample answer: The *solubility* of *sugar* in water increases with *temperature*. At higher temperatures, more *solute* may be added without *saturating* the *solution*.

Correction/Completion

2. Answer:
   a. Acid rain occurs when *sulfur* dioxide dissolves in rainwater, which falls through the atmosphere.
   b. *Solubility* is the measure of the amount of solute needed to make a solution saturated.
   c. When maple syrup is made from maple sap, a solution is made more *concentrated*.

Short Response

3. Answer:

a.	Concentrated	water from the Dead Sea
	Dilute	water from the ocean
b.	Dilute	nectar
	Concentrated	honey
c.	Dilute	maple sap
	Concentrated	maple butter
d.	Concentrated	carbon dioxide in car exhaust
	Dilute	carbon dioxide in the atmosphere

Numerical Problem

4. Answer:

$$\frac{3.5\ g}{100\ g} \times 40\ L \times \frac{1000\ g}{1\ L} = 1400\ g \text{ of salt to } 40\ L \text{ of water}$$

Data for Interpretation

5. Answer:
   If there are 2.2 g of baking soda per milliliter, and there are 5 mL per teaspoon, then there are 2.2 g X 5, or 11 g, of baking soda per teaspoon. The solubility of baking soda at 20°C is 9.6 g per 100 g of water, or 9.6 g per 100 mL of water. Therefore, the teaspoon of baking soda will not dissolve in the 100 mL of water that Friedrich is using. He will need to heat the water to 30°C before all of the baking soda will dissolve into solution so that he can make his cake.

   Difficulty: 3

Short Essay

6. Answer:
   Sample answer: Salt lowers the freezing temperature of water. If enough salt is added to ice (frozen water) on roads, it will lower the freezing point to below the temperature of the ice, so the ice will melt. In a container, such as a bucket, adding salt to ice will reduce the temperature of the contents, so the container will become colder. A second container that has a lower salt concentration than the ice-salt mixture in the first container can be immersed in the first container, and the contents in the second container will freeze.

   Difficulty: 4

Unit 3 End-of-Unit Assessment

Word Usage

1. a. Use all of the following terms in one or two sentences to show how they are related: precipitation, evaporation, vapor, cooling, sun, rises, condensation, water cycle.

   b. Barbara gradually added salt to a beaker of water until no more salt would dissolve. Use all of the following words to describe the solution at different times: dilute, concentrated, saturated, and unsaturated.

Correction/Completion

2. The definitions and explanations below were given by students. Check their work by correcting any mistakes you find or by giving a better definition or explanation.
   a. Lisa's definition of solubility: "The solubility of something is how much of it can be dissolved in water."

   b. Melea's explanation of how to make maple syrup: "You heat up the tree sap until it gets really thick."

   c. Nathan's definition of hard water: "Hard water is water that doesn't make suds easily."

   d. Tameka's explanation of how to tell a dilute solution from a concentrated solution made from the same solute and solvent: "You can compare the weights."

Short Response

3. How does a solution differ from a non-solution?

4. a. What effect does adding salt have on the freezing point of water?

   b. What effect does adding salt have on the boiling point of water?

   c. Give an example of when and why it is helpful to add salt to ice.

5. Honey and maple syrup are both thick, sweet solutions, but the processes for making each are very different. Describe the differences and similarities between the processes for making these sugary solutions.

6. For each situation, name the solute and the solvent. Then identify whether each part of the solution is a gas, a solid, or a liquid. Record your answers in the table below.

Situation 1: Mariana always likes her tea sweetened.
Situation 2: Lee opened a bottle of cologne. Before long, the smell of perfume was everywhere.
Situation 3: The grass stain was easily removed with alcohol.

Situation	Solute	Solid, liquid, or gas	Solvent	Solid, liquid, or gas
1				
2				
3				

Illustrative

7. In the test tubes below are three different concentrations of the same solution. Complete the drawings by sketching a pencil in the correct locations in test tubes *B* and *C*.

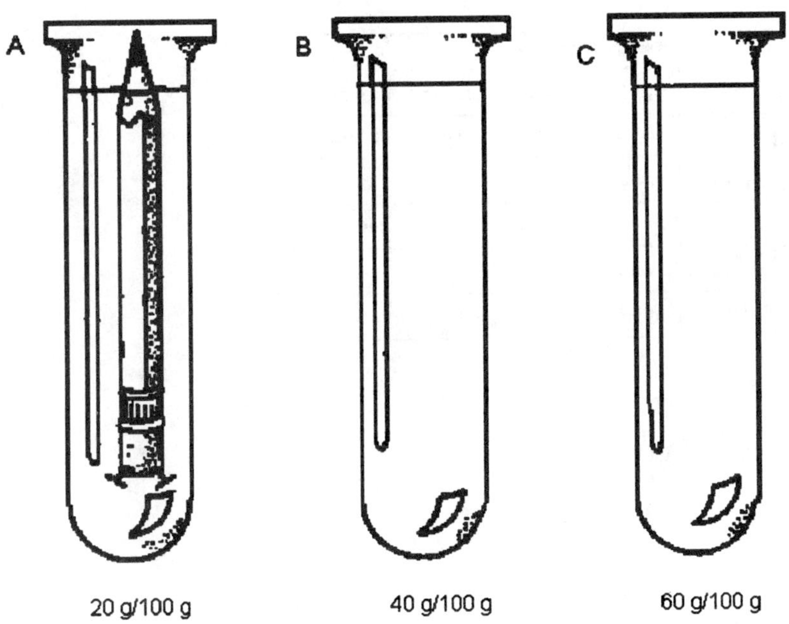

Data for Interpretation

8. The table shows the maximum number of grams of different solutes that can be dissolved in 100 g of water at different temperatures. Use this information to answer the questions that follow.

   a. Which is more soluble at 0°C, table salt or potassium chloride?

   ____

   b. Of all of the substances, which substance is the most soluble at 50°C?

   ____

   c. What effect does an increase in temperature have on solubility?

   ____

   ____

   ____

   ____

	Temperature					
Substance	0°C	10°C	20°C	30°C	40°C	50°C
table salt (solid)	35.7 g	35.8 g	36.0g	36.3 g	36.6g	37.0 g
potassium chloride (solid)	27.6 g	31.0 g	34.0g	37.0g	40.0g	42.6 g
sugar (solid)	179.0 g	191.0 g	204.0g	220.0g	238.0g	260.0 g
sulfur dioxide (gas)	22.8 g	16.2 g	11.3g	7.8 g	5.4g	4.5 g
oxygen (gas)	0.0070g	0.0054g	0.0044 g	0.0037 g	0.0033g	0.0030g

Numerical Problem

9. Kim wanted to know how much sugar she could dissolve in 50 mL of water.
   a. What was the concentration of the sugar solution when she had dissolved 50 g of sugar?

   b. When the temperature was 20°C, she was able to dissolve 100 g of sugar in 50 mL of water. What is the solubility of sugar in water at 20°C?

   c. It takes 2 kJ to evaporate 1 g of water. How much energy is required for Kim to recover all of the sugar dissolved in her solution?

Short Essay

10. Arturo loves canned peaches, but he is on a diet. He wants to know which brand of canned peaches uses the lowest concentration of sugar in the syrup. Unfortunately, he cannot taste or see any difference. Suppose that he has a mass scale, clear straws, food coloring, an electric stove, Pyrex bowls, a thermometer, and a refrigerator. Explain to him in detail two or three different ways he could find out which syrup has the lowest concentration of sugar.

11. Suppose that you and your family are stranded on a deserted beach and the family car will not start. Beyond the beach is hot desert. The only source of water is the ocean. There is dry wood on the beach, and you have matches, cooking and eating utensils, and a funnel and hose from the car. Explain how you could obtain drinking water.

Graphic

12. Use the graph below, which shows how the solubility of three different salts changes with temperature, to answer the questions that follow.

a. Which substance (*A*, *B*, or *C*) is the most soluble at 0°C? What is its solubility at that temperature?

____

____

b. Which substance is the most soluble at 80°C? What is its solubility at that temperature?

____

____

c. At what temperature is the solubility of substance *A* the same as that of substance *C*? What is their solubility at that temperature?

____

____

d. Make a general statement about the effects of temperature on solubility.

____

____

____

____

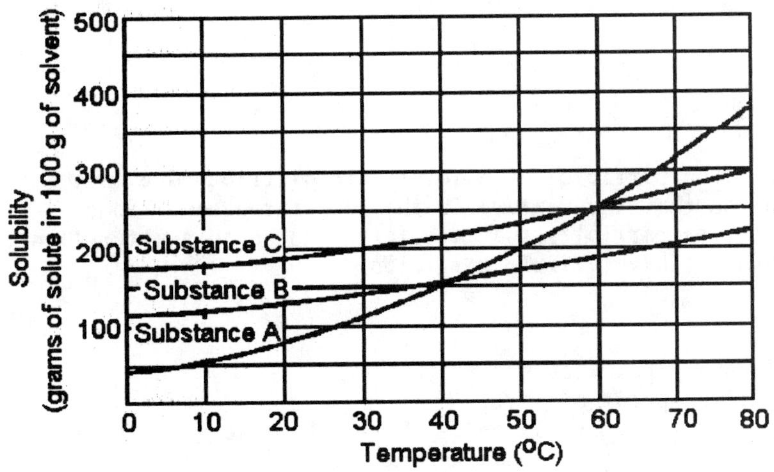

Answers to Unit 3 End-of-Unit Assessment

Word Usage

1. Answer:
   a. Sample answer: In the *water cycle*, heat from the *sun* causes *evaporation* of water on the Earth's surface. The water *vapor rises* and cools. As a result of *cooling*, the gas changes back into a liquid in a process called *condensation*, and the water falls back to the Earth in the form of *precipitation*.
   b. Sample answer: When Barbara first adds a little salt to the beaker of water, her solution is *dilute*. The more salt she adds, the more *concentrated* the solution becomes. As long as more salt can dissolve in the water, her solution is *unsaturated*. When no more salt can dissolve in the water, the solution is *saturated*.

Correction/Completion

2. Answer:
   a. Sample answer: Solubility is a measure of the amount of solute that can dissolve in a given amount of solvent under given conditions.
   b. Sample answer: Maple syrup is made by boiling sap until it has the proper sugar and water content. As the water boils off and the sugar becomes more concentrated, the solution thickens. This occurs at a specific temperature that is over 100°C.
   c. Sample answer: Hard water is water that contains dissolved solutes. These dissolved solutes make forming suds difficult.
   d. Sample answer: You can compare the masses of equal amounts of two solutions made from the same solute with identical amounts of solvent to determine which one is more concentrated. The one with the greater mass is more concentrated because it contains more solute.

Short Response

3. Answer:
   Sample answer: In a solution, the particles of a solute are distributed evenly among the particles of the solvent so that the mixture is the same throughout. Some non-solutions contain undissolved particles that are large enough to scatter light. Other non-solutions are cloudy or have visible particles.

4. Answer:
   a. Adding salt lowers the freezing point of water.
   b. Adding salt raises the boiling point of water.
   c. Possible answers: Adding salt to icy roads melts the ice because salt lowers the freezing point of water. This makes the roads safer to travel. Adding salt to ice used in ice-cream making is helpful because it lowers the temperature of the melted water surrounding the ice cream to about -5°C.

5. Answer:
   Sample answer: Honeybees use evaporation by fanning their wings in front of the cells containing nectar to remove excess water and concentrate the sugar. In making maple syrup, however, excess water is removed by boiling. This rapidly changes the liquid into a gas.

6. Answer:

Situation	Solute	Solid, liquid, or gas	Solvent	Solid, liquid, or gas
1	sugar, tea	solid	water	liquid
2	scent, alcohol	gas, liquid	air	gas
3	grass stain	solid	alcohol	liquid

Illustrative

7. Answer:

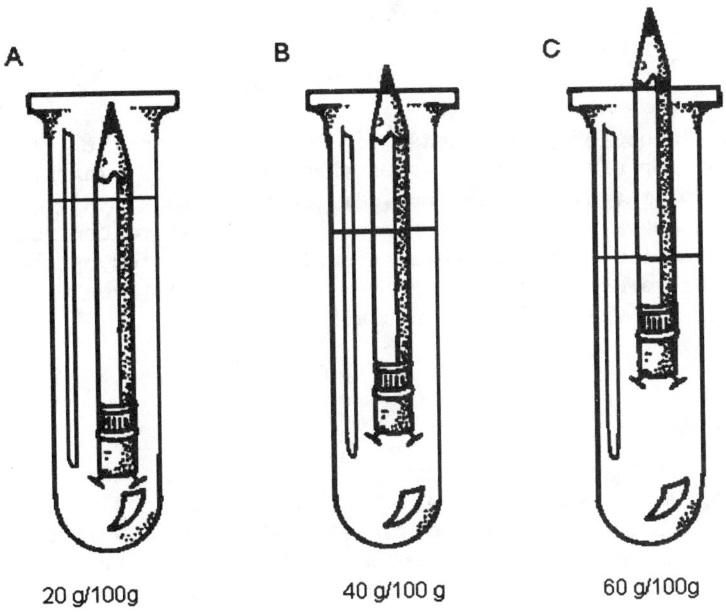

Data for Interpretation

8. Answer:
   a. Table salt
   b. Sugar
   c. An increase in temperature increases the solubility of a solid but decreases the solubility of a gas.

Numerical Problem

9. Answer:
   a. Concentration is often expressed in terms of grams of solute per 100 g of solvent. Kim's solution has 50 g of sugar in 50 mL of water (about the same as 50 g of water), so the concentration of her solution is 100 g of sugar per 100 g of water.
   b. The solubility of sugar in water at 20°C is 200 g of sugar per 100 g of water.
   c. 50 mL of water is equivalent to 50g. 50 g x $\frac{2kJ}{1g}$ = 100 kJ of energy

   Difficulty: 3

Short Essay

10. Answer:
    Each method of comparison applies to one of these ideas: (1) The syrup with the lowest concentration of sugar will sit on top of the other syrups when layered. (2) In the case of equal volumes of each syrup, the syrup with the least amount of sugar will have the lowest mass. (3) The boiling temperature of the syrup with the lowest concentration of sugar will be the lowest. An acceptable answer should clearly describe two of these ways to identify the syrup with the lowest concentration of sugar. If all three ways are described clearly and in detail, give the student a bonus.

    Difficulty: 3

11. Answer:
    Answers should apply the following basic ideas: Pure water can be obtained by distillation. This can be done by using the matches, dry wood, and a pot to boil the water, and then by using the funnel to pass the steam through a hose, where it can cool and condense. The salt is left behind in the first pot, and the water can be collected in a second pot.

Graphic

12. Answer:
    a. Substance *C* is the most soluble at 0°C, with a solubility of 170 g per 100 g of solvent.
    b. Substance *A* is the most soluble at 80°C, with a solubility of 430 g per 100 g of solvent.
    c. Substances *A* and *C* both have a solubility of 250 g per 100 g of solvent at 60°C.
    d. Sample answer: The solubility of some substances is greatly affected by temperature, as shown with substance *A*. Its solubility increases a great deal when the temperature increases. Other substances are not affected as much, as with substances *B* and *C*, whose solubilities do not increase as much when the temperature increases.

Unit 3 Activity Assessment

Activity Assessment

1. **Solution or Not?**

Teacher's Notes

Overview
Students test four different substances to determine if they are solutions or non-solutions.

Materials
(per activity station)
- a flashlight
- four 500 mL beakers
- orange gelatin
- orange-flavored drink-mix powder
- pulpy orange juice
- orange soda

Preparation
The day before the activity, prepare the gelatin and the drink mix according to the instructions given on the packages. Prior to the assessment, equip student activity stations with the materials needed for each experiment. Remind students not to taste any of the substances at their activity stations.

Time Required
Each student should have 15 minutes at the activity station and 10 minutes to complete the Identification Chart.

Performance
At the end of the assessment, students should turn in the following:
- a completed Identification Chart

Evaluation
The following is a recommended breakdown for evaluation of this Activity Assessment:
- 10% appropriate use of materials and equipment
- 30% demonstration of knowledge of the Tyndall effect
- 30% clarity and quality of observations
- 30% ability to form a logical conclusion based on observations

2. **Solution or Not?**
"Knock-knock." "Who's there?" "Orange." "Orange who?" "Orange you wondering which of us is a solution?" All kidding aside, you will find four similar substances at your activity station. Some of them are solutions—and some of them are not. Using what you know about the properties of solutions, determine whether each substance is a solution.

Before You Begin . . .
As you work through the tasks, keep in mind that your teacher will be observing:
- how you use the materials and equipment
- the quality and clarity of your observations
- if you correctly identify the substances as solutions or non-solutions

Now you're ready to find the solutions!

Task 1 Using scientific skills, determine whether each substance is a solution or a mixture. Remember never to taste any of the substances that you are testing. Record your observations and identification of each in the Identification Chart.

Task 2 For the substances that are solutions, identify the solute(s) and the solvent. For the mixtures, identify what you think the different parts of the mixture are.

Task 3 One of these substances is a *colloid.* A colloid is a mixture that has extremely tiny, undissolved particles—particles so tiny that you cannot easily see the individual particles that make up the mixture. These particles are bigger than the ones found in solutions, but smaller than the particles seen in most mixtures. Indicate which substance you think might be a colloid.

3. **Identification Chart**

Substance	Observations	Identification—solution or mixture
orange juice		
orange soda		
orange gelatin		
orange-flavored drink mix		

Answers to Unit 3 Activity Assessment

1. Answer: Not applicable (teacher's notes)

2. Answer: Not applicable (student's notes)

3. Answer:

Identification Chart

Substance	**Observations**	**Identification—solution or mixture**
orange juice	Tyndall effect; parts: juice, pulp (Some students may say concentrate and water.)	mixture
orange soda	no Tyndall effect; solvent: water; solutes: orange flavoring, carbon dioxide, orange coloring, sugar	solution
orange gelatin	Tyndall effect; parts: water, gelatin, air bubbles	mixture—colloid
orange-flavored drink mix	no Tyndall effect; solvent: water; solute: drink-mix powder	solution

Unit 3 SourceBook Assessment

Multiple Choice

1. Two or more substances that can be separated by physical means are called a
   a. solution. b. suspension. c. colloid. d. mixture.

2. A cup of coffee is an example of a
   a. solution. b. colloid. c. suspension.

3. In every solution, there is (more/less) solute than solvent. (Circle the correct word.)

4. Bottles whose labels read "Shake well before using" most likely contain a
   a. solution. b. colloid. c. suspension.

5. Gelatin is an example of a
   a. solution. b. colloid. c. suspension.

6. Which of the following is not an acid?
   a. soap b. aspirin c. orange juice d. cola

7. The (more/fewer) hydronium ions there are in a solution, the stronger an acid it is. (Circle the correct word.)

Correction/Completion

8. Put the correct name under each of the following chemical groupings.
   H_3O^+ + OH^- → $2H_2O$
   __________ + __________ → __________

Multiple Choice

9. The chemical reaction $H_3O^+ + OH^- \rightarrow 2H_2O$ illustrates the process known as
   a. equilibrium. b. deionization. c. neutralization. d. dissolution.

10. Heartburn remedies are an example of a(n)
    a. acid. b. base.

11. Orange juice would most likely have a pH of
    a. 3.0. b. 7.0. c. 10.0.

12. Laundry detergent would most likely have a pH of
    a. 3.0. b. 7.0. c. 10.0.

13. The contents of your stomach are
    a. acidic. b. basic.

Correction/Completion

14. When an acid and a base are neutralized, the reaction forms _________ and _________.

Multiple Choice

15. Why do people who live in cold climates pour salt on their icy driveways in winter?
    a. to lower the temperature at which water freezes
    b. to provide traction for the wheels on their car
    c. to neutralize the ice
    d. to make the pH of the ice more acidic

Matching

16. Match the solutions on the left to the correct descriptions on the right.

a. dilute solution	___ has all of the solute it can hold without changing any conditions
b. concentrated solution	___ has a small amount of solute in a large amount of solvent
c. saturated solution	___ has a large amount of solute in a small amount of solvent

17. Match the substances on the left with the correct descriptions on the right.

	___ bitter taste
	___ sour taste
	___ soapy feel
a. acid	___ NaOH
	___ automobile batteries
b. base	___ HCl
	___ swimming pools
	___ alkaline
	___ sour milk
	___ cement

True/False

18. In every solution, either the solute or the solvent must be a liquid.
    a. true b. false

19. Unlike acids, all bases are safe to touch.
    a. true b. false

Short Response

20. You are making a pitcher of tea, and you put 6 tablespoons of sugar in the water, but not all of the sugar dissolves. Describe one way to dissolve the sugar.

____

____

____

____

____

21. Terry accidentally poured a cup of lemon-lime soda instead of a cup of water into the fishbowl. The two goldfish in the fishbowl soon died. What probably killed the fish?

____

____

____

____

____

22. Why is it important that your blood's pH change by no more than 0.4 (up or down the pH scale)?

____

____

____

____

____

Correction/Completion

23. The pH scale ranges from ______ to ______.

24. Place the ions on the left with the descriptions on the right.

a. hydroxide ion	______ H_3O^+	______ found in bases
b. hydronium ion	______ OH^-	______ found in acids

True/False

25. All salts are white.
a. true b. false

Answers to Unit 3 SourceBook Assessment

Multiple Choice

1. Answer: d. mixture.

2. Answer: a. solution.

3. Answer: less

4. Answer: c. suspension.

5. Answer: b. colloid.

6. Answer: a. soap

7. Answer: more

Correction/Completion

8. Answer: Acid + Base → Water

Multiple Choice

9. Answer: c. neutralization.

10. Answer: b. base.

11. Answer: a. 3.0.

12. Answer: c. 10.0.

13. Answer: a. acidic.

Correction/Completion

14. Answer: water and salt

Multiple Choice

15. Answer: a. to lower the temperature at which water freezes

Matching

16. Answer:

c has all of the solute it can hold without changing any conditions
a has a small amount of solute in a large amount of solvent
b has a large amount of solute in a small amount of solvent

17. Answer:

b bitter taste
a sour taste
b soapy feel
b NaOH
a automobile batteries
a HCl
a swimming pools
b alkaline
a sour milk
b cement

True/False

18. Answer: b. false

19. Answer: b. false

Short Response

20. Answer:
Sample answer: Heating the water would dissolve the remaining sugar, or adding more water to the mixture might dilute the solution enough so that the remaining sugar would dissolve.

21. Answer:
The lemon-lime soda made the water in the fishbowl more acidic. The change in acidity probably killed the fish.

22. Answer:
If the pH exceeds this limit, the body will not function properly. If the change is too great, or if it is not corrected rapidly, it can be fatal.

Correction/Completion

23. Answer: The pH scale ranges from 0 to 14.

24. Answer:

a. hydroxide ion b	H_3O^+	a found in bases
b. hydronium ion a	OH^-	b found in acids

True/False

25. Answer: b. false

Unit 3 Extra Assessment Items

Word Usage

1. Whitney's group experimented to find out what helps the most to speed up the process of dissolving candy. Use as many of the following words as you can to describe what they did: variable, constant, observations, prediction, inference, hypothesis, and conclusion.

Correction/Completion

2. The definitions and explanations below were given by students. Check their work by correcting any mistakes you find or by giving a better definition or explanation.
   a. Cara's explanation of how to tell a solution from a non-solution: "You can tell a solution because it is clear."

   ____
   ____
   ____
   ____

   b. Scott's definition of a saturated solution: "When you can't put any more stuff in something, it's saturated."

   ____
   ____
   ____
   ____

   c. Bianca's explanation of the water cycle: "Water is warmed by the sun and evaporates into the air. In the air, it cools and condenses to form drops. The drops fall as rain or snow."

   ____
   ____
   ____
   ____

   d. Cory's explanation of the difference between boiling and evaporation: "Boiling and evaporation are the same, except boiling is hotter."

   ____
   ____
   ____
   ____

Short Essay

3. How could you find out whether a liquid is a solution or a non-solution? Provide a written description and an illustration of an experiment you could use.

____

____

____

____

____

____

____

____

____

Illustration

4. How could you use paper chromatography to find out the difference between the inks used in different pens? Provide a written description and an illustration of an experiment you could use.

____

____

____

____

____

____

Illustration

5. How does the change in the boiling point of a sugar solution change as the concentration of sugar changes? Provide a written description of an experiment you could use to find out.

____

____

____

____

____

____

____

____

____

____

Short Response

6. Name the solute and solvent (if any) in each case.
   a. a carbonated beverage

   b. a mixture of 10 mL of cherry-flavored syrup and 50 mL of cola

   c. an alloy of 10 percent tin, 30 percent zinc, and 60 percent copper

   d. pure methyl alcohol

   e. maple syrup

   f. the ocean

   g. oil and water

7. Give an example of each of the following kinds of solutions.
   a. The solute is a gas and the solvent is a liquid.

   ____

   b. The solute is a solid and the solvent is a liquid.

   ____

   c. The solute and the solvent are both gases.

   ____

   d. The solute and the solvent are both liquids.

   ____

   e. The solute and the solvent are both solids.

   ____

8. For each situation given, name the solute and the solvent. Then identify whether each part of the solution is a gas, a solid, or a liquid. Record your answers in the table provided.
   Situation 1: Jared spilled oil on the floor. He wiped it up with a cloth moistened with mineral spirits.
   Situation 2: When a glass of cold water warms up, little bubbles appear on the inside of the glass.
   Situation 3: Iron melts at 1500°C, and then some copper is added. When the mixture cools, you get bronze, an alloy.
   Situation 4: A little food coloring added to carbonated water made it sparkling red.
   Situation 5: A carbonated beverage is perfectly clear; when it is opened, a lot of fizzing occurs.

Situation	Solute	Solid, liquid, or gas	Solvent	Solid, liquid, or gas
1				
2				
3				
4				
5				

9. You are a salesperson for a product that softens water. You wish to convince your customer that he or she should purchase your product. What could you do or say to convince the customer that his or her water is hard?

   ____

   ____

   ____

   ____

   ____

10. Suppose that you have collected samples of water from five different wells in your community. Explain in detail how you would conduct a fair test to see which well contains the hardest water and which contains the softest.

Graphic

11. The following graph shows the solubility of three different compounds called salts: silver nitrate ($AgNO_3$), sodium nitrate ($NaNO_3$), and potassium nitrate (KNO_3).

    a. Which compound is the most soluble at 0°C?

    ____

    b. Which compound is the least soluble at 0°C?

    ____

    c. Which compound is the least soluble at 80°C?

    ____

    d. How much potassium nitrate can be dissolved in 100 g of water at 50°C?

    ____

    e. For which of the salts shown does the solubility increase the most between 0°C and 20°C?

    ____

    f. What two factors increase the solubility of the salts?

    ____

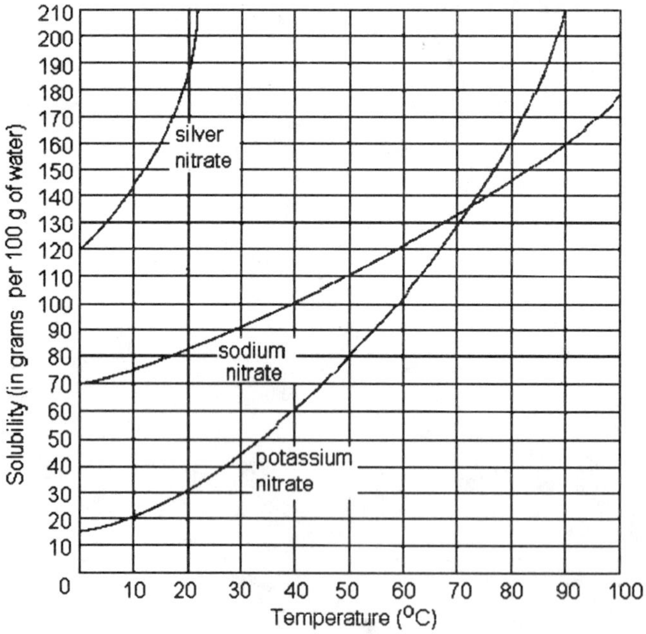

Illustrative

12. You are given four bottles filled with liquid, as shown below.

    a. **Without further information**, which can you say for sure are solutions and which are not?

    ____

    ____

    ____

    ____

    ____

    ____

    b. What could you do to find out for certain whether the liquids are solutions?

    ____

    ____

    ____

    ____

    ____

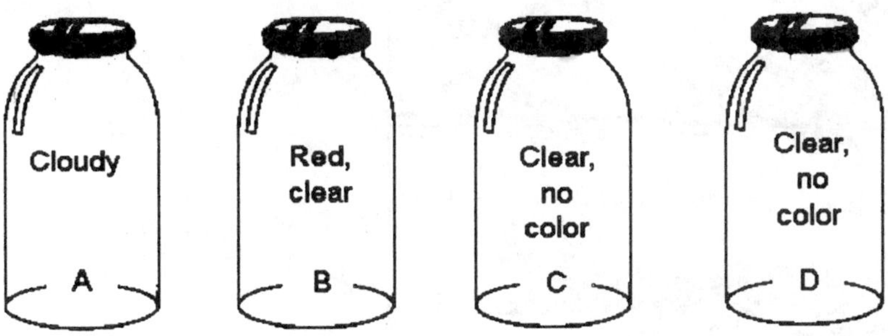

13. Identify the solutions in the illustration. In each case, name the solvent and any solutes that you think may be present.

Performance Task

14. **Materials**
    - 5 mixtures in glass beakers, labeled *A*, *B*, *C*, *D*, and *E*
    - a flashlight
    - an index card

    Task
    Determine which of the mixtures are solutions, and state the reasons for your conclusion in each case.

Mixture	Solution or non-solution	Explanation
A		
B		
C		
D		
E		

15. **Materials**
    - 2 solutions of sugar and water in beakers
    - an empty beaker
    - 2 test tubes
    - a graduated cylinder
    - a clear drinking straw or glass tube
    - an equal-arm balance
    - a pencil hydrometer (pencil with a tack in the eraser)

    Task
    Determine which of the solutions is the most concentrated. Check your results by using another method to determine concentration. Record your data and observations and state the reasons for your conclusion on a separate piece of paper.

16. **Materials**
- sugar
- a graduated cylinder
- a thermometer
- hot and cold water
- test tubes
- a stirring rod
- an equal-arm balance

Task

For each of three different water temperatures, find out how much sugar you can dissolve in 10 mL of water. Record your results in the table. Then plot your results in a graph of solubility (grams per 100 mL) versus temperature. Remember to label the *x*- and *y*-axes.

	Amount of sugar dissolved in 10 mL of water
Temperature 1: ______	
Temperature 2: ______	
Temperature 3: ______	

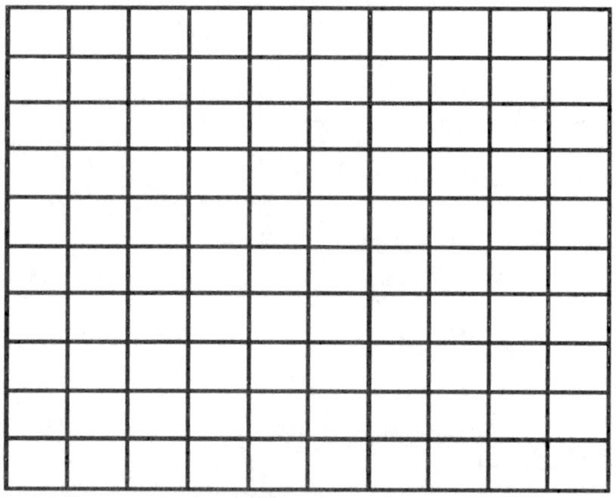

17. It is difficult to write good test questions. You have volunteered to help your teacher by writing test questions to meet the following criteria:
   a. Write a story of approximately 100 words that includes five words describing solutions. Students will be asked to pick out the solution words in the story. Assume that your copy is for the teacher, and underline all of the solution words that you have used.
   b. Draw an illustration showing solutions and non-solutions that can be found in the kitchen. Students will be asked to identify the solutions and name the solvent and solutes. Provide your answers for the teacher in a chart with the following headings: Solutions, Solvents, and Solutes.

18. As part of an environmental task force, you have volunteered to do the following:
   a. identify a major source of air, water, or soil pollution
   b. find out how this problem might be solved

   With two or three other students, carry out this investigation. Present your ideas and proposals in the form of a poster. Use solution words to help you explain your ideas.

Answers to Unit 3 Extra Assessment Items

Word Usage

1. Answer:
   Sample answer: To find out what helps the most to speed up the dissolving of candy, Whitney's group first made the *inference* that heating and stirring are the best ways to speed up the rate of dissolving, based on previous *observations*. They formed the *hypothesis* that heating is the fastest way to dissolve candy in water. To test each *variable*, they placed water and a piece of candy into each of two beakers. Since all factors except the ones to be tested should be held *constant* for accurate results, they used beakers of the same size, the same kind and amount of candy, and the same amount of water in each beaker. One beaker was stirred until all of the candy dissolved, while the other beaker was heated until all of the candy dissolved. The time it took for the candy to completely disappear was measured. Since the candy took less time to dissolve with stirring than with heating, the group came to the *conclusion* that stirring, not heating, is the fastest way to dissolve candy in water. They also made the *prediction* that heating and stirring at the same time would make candy dissolve even faster than stirring alone.

Correction/Completion

2. Answer:
   a. Sample answer: You can tell a liquid solution from a liquid non-solution because the solution does not exhibit the Tyndall effect. That is, you cannot see the path of light through a solution.
   b. Sample answer: When no more solute can be dissolved in the solvent under given conditions, a solution is saturated.
   c. Sample answer: Water from oceans, lakes, rivers, and land is warmed by the sun and evaporates, forming water vapor. In the air, the water vapor cools and condenses to form droplets of liquid water. When these droplets grow large enough, they eventually fall as precipitation.
   d. Sample answer: Boiling and evaporation involve the same process, but in boiling, this process is sped up.

Short Essay

3. Answer:
   Sample answer: If necessary, the mixture should be stirred and time should be allowed for any undissolved materials to settle. If the liquid is still murky, it probably contains suspended particles and is therefore not a solution. On the other hand, if the liquid appears to be clear, it is still a good idea to check for the Tyndall effect. Do so by passing a beam of light through the liquid. If the beam can be seen within the liquid, it must contain suspended particles that are reflecting light. It is therefore not a solution. If the beam cannot be seen, it probably is a solution.

   Illustration
   Illustrations will vary but should clearly demonstrate a method for determining how to tell a solution from a non-solution. For example, students may choose to show how to test for the Tyndall effect by using a beaker, a flashlight, and an index card.

4. Answer:
As shown in Chapter 8, Exploration 2, on page 177 of your textbook, different patterns of colors result from different inks when tested by paper chromatography.

Illustration
Illustrations should clearly demonstrate the process of paper chromatography. For example, see Chapter 8, Exploration 2, on page 177 of your textbook.

5. Answer:
Sample answer: As the concentration of sugar in a sugar solution increases, the boiling point of the sugar solution also increases. This can be tested by heating sugar solutions with different concentrations of sugar and measuring the boiling points of each.

Short Response

6. Answer:
a. Solutes: carbon dioxide, sugar, and flavoring; students may also list ingredients such as caffeine, food coloring, and phosphoric acid; solvent: water
b. Solutes: sugar, cherry flavoring, cola flavoring, caffeine, carbon dioxide; solvent: water
c. Solutes: zinc and tin; solvent: copper
d. Solutes: none; solvent: none. Methyl alcohol is a single substance.
e. Solutes: sugar, maple flavoring; solvent: water
f. Solutes: salt and compounds of magnesium, sulfur, calcium, potassium, bromine, and iodine; solvent: water
g. Solutes: none; solvents: none. Oil and water don't mix.

7. Answer:
a. Sample answers: carbonated water
b. Sample answers: Baking soda and water; syrup; salt water; the ocean
c. Sample answers: The atmosphere; gases emitted from factories
d. Sample answers: Vinegar; antifreeze; mineral spirits
e. Sample answers: 10 karat, 14 karat, or 18 karat gold; brass; steel; all metal alloys

8. Answer:

Situation	Solute	Solid, liquid, or gas	Solvent	Solid, liquid, or gas
1	oil	liquid	mineral spirits	liquid
2	air	gas	water	liquid
3	copper	solid*	iron	solid*
4	food coloring, CO_2	liquid, gas	water	liquid
5	carbon dioxide	gas	water	liquid

* Students may correctly identify the metal as liquid when it is heated. Accept both answers.

9. Answer:
Sample answer: I could perform the "soap test." This involves demonstrating that it is more difficult to form suds in the hard water than it is to form suds in the water with the softener added.

10. Answer:
Sample answer: In a fair test, the water temperature and the amounts of water, soap, and stirring must be the same for each sample. Consider the answer acceptable if three of these variables are kept the same. If all four of these variables are kept the same, give the answer a bonus.

Graphic

11. Answer:
   a. Silver nitrate
   b. Potassium nitrate
   c. Sodium nitrate
   d. About 80 g
   e. Silver nitrate
   f. Temperature, concentration

Illustrative

12. Answer:
   a. Bottle *B* most likely contains a solution. Since all common solvents are colorless, some other dissolved substance is causing the red color. Bottles *C* and *D* could be pure because there are no visible clues that indicate the presence of another substance. Bottle *A* contains undissolved particles that are making the mixture cloudy, so it cannot be a solution.
   b. You can boil the liquids in bottles *B*, *C*, and *D* to see if any solute is left behind. You could check for the Tyndall effect and eliminate any liquid that shows the path of light shining through it.

13. Answer:
Sample answer: Ocean: Solutes include salts and compounds of magnesium, sulfur, calcium, potassium, bromine, and iodine. The solvent is water.
Smoke from factory: Solutes include sulfur dioxide, carbon dioxide, and carbon monoxide. The solvent is air, which is composed mostly of nitrogen.
Atmosphere: Solutes include oxygen, hydrogen, carbon dioxide, argon, neon, and helium. The main solvent is nitrogen.
Some of the buildings and the boat probably contain alloys—solutions made of two or more solids.
Some students might note that the liquid inside leaves is a solution made of water (the solvent) and salts, proteins, and sugars (the solutes).

Performance Task

14. Answer:

Mixture	Solution or non-solution	Explanation
A		Answers will vary depending on the mixtures provided but should be clear and accurate.
B		Student answers should reflect an understanding of the Tyndall effect.
C		
D		
E		

15. Answer:
Answers will vary depending on solutions provided but should be clear and accurate. Data, observations, and conclusions should be clearly outlined. Acceptable methods include, but are not limited to, the following: using a pencil as a hydrometer, finding the mass of equal-volume samples, or evaporating equal-volume samples and finding the mass of the residues. Some students may be tempted to try a taste test. The method should be discouraged before students begin the assessment since tasting unknown substances is never a good idea in the laboratory.

16. Answer:
Answers will vary but should demonstrate the ability to carefully keep track of data and to express that data on a graph.

17. Answer: Answers will vary but should be clear and logical. Encourage creativity.

18. Answer:
Answers will vary but should demonstrate the student's ability to recognize a genuine environmental problem and to seek a logical solution to that problem. Posters should be clear and effective and should demonstrate a real-world application of the concepts learned in this unit.

Chapter 10 Assessment

Word Usage

1. Describe a tennis player serving a ball. In doing so, use all of the following words: force, agent, receiver, and effect.

Correction/Completion

The statements below are incorrect or incomplete. Your challenge is to make them correct and complete.

2. A block of wood has the same weight everywhere in the universe. However, its mass in space is less than its mass on Earth.

3. A balloon that has been rubbed by a cloth places a gravitational force on a small piece of paper. This is an example of a contact force.

4. Complete the following chart to show what happens to an object when a force is applied to it.

Forces on a stationary object	Result	Forces on a moving object	Result
 Balanced		 Balanced	
Unbalanced	Object begins moving.	 Unbalanced	

Short Response

5. If you were a defensive linebacker for a football team, which sort of ball carrier would be easier to tackle?
   a. a large, slow runner
   b. a large, fast runner
   c. a small, slow runner
   d. a small, fast runner

   Explain your choice.

Short Essay

6. Imagine that a spaceship is traveling from the Earth to the moon. When it is exactly halfway between the two bodies, is it influenced more by the pull of the Earth's gravity or by the pull of the moon's gravity? Explain your answer.

Answers to Chapter 10 Assessment

Word Usage

1. Answer:
   Sample answer: When the player's racket strikes the ball, it exerts a pushing *force*. The racket is the *agent* of the force, and the ball is the *receiver* of the force. The *effect* of the force is that the ball is set in motion in the direction of the force.

Correction/Completion

2. Answer:
   A block of wood has the same *mass* everywhere in the universe. However, its *weight* in space is less than its *weight* on Earth.

3. Answer:
   A balloon that has been rubbed by a cloth places an *electrical* force on a small piece of paper. This is an example of a *noncontact* force.

4. Answer:

Forces on a stationary object	Result	Forces on a moving object	Result
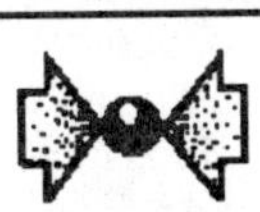 Balanced	Object remains stationary.	 Balanced	Object remains at constant speed or velocity.
 Unbalanced	Object begins moving.	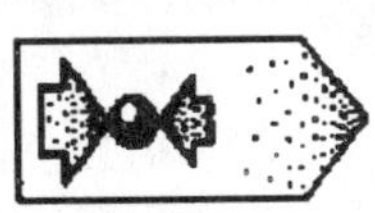 Unbalanced	Object increases speed or velocity.

Short Response

5. Answer:
   The small, slow runner (c) has the least mass and the least speed, so he or she would require the least force to tackle. As a result, the small, slow runner would be easier to tackle.

   Difficulty: 3

Short Essay

6. Answer:
Sample answer: The spaceship is influenced more by the pull of the Earth's gravity. Some students may state that the gravitational pulls are equal because the distances are equal. However, the gravitational pull between two objects is influenced by the mass of the objects as well as by the distance between them. Since the Earth's mass is much greater than that of the moon, the Earth's pull is greater at the halfway point.

Difficulty: 4

Chapter 11 Assessment

Word Usage

1. Use all of the following terms in one or two sentences to show how the terms are related: newton, kilogram, mass, force, and weight.

____

____

____

____

Correction/Completion

2. The statements below are incorrect or incomplete. Your challenge is to make them correct and complete.
   a. No matter where you go in the universe, your weight measured in newtons will be the same.

   ____

   ____

   b. On Earth, a mass of 50 kg weighs approximately 5 N.

   ____

   ____

Short Response

3. Which is the best tool for measuring each of the following?

   a. a person's weight on Earth ________________

   b. the mass of a bowling ball on Earth ________________

   c. the mass of a bowling ball on the moon ________________

Illustrative

4. Label the arrows on the diagram below to show the points where pushing down to lift the weight would require 10 N, 50 N, and 100 N of force.

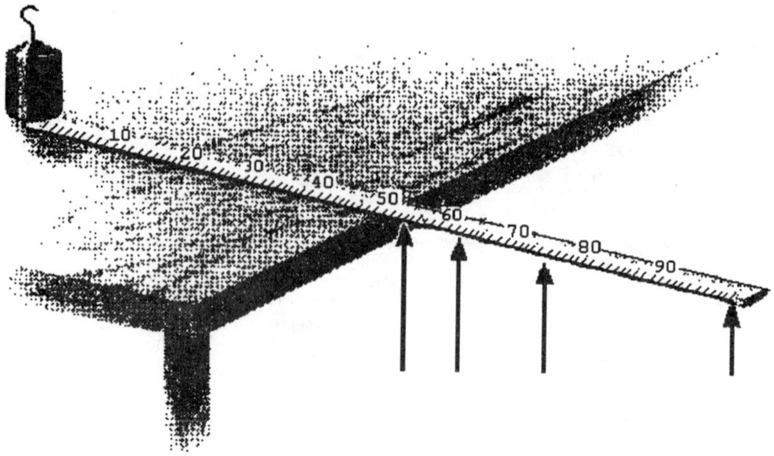

Numerical Problem

5. Ella made a simple force meter with a large rubber band. She found that the weight of 10 large marbles stretched her rubber band 2 cm. Using a commercial force meter, she found that her 10 marbles weighed almost exactly 1 N.
   a. What can Ella conclude about the mass of each marble?

   ____

   ____

   b. What can she conclude if the weight of an object stretches her rubber band 10 cm?

   ____

   ____

Short Response

6. You have been given a 1 kg cube of cheese and have been asked to slice a piece with a mass of exactly 125 g. The only tools you have are a knife, a string, and an equal-arm balance. What method could you use to ensure that you end up with 125 g of cheese?

   ____

   ____

   ____

   ____

   ____

Answers to Chapter 11 Assessment

Word Usage

1. Answer:
   Sample answer: A *newton* is a measure of the gravitational *force* exerted on an object, which gives the object its *weight*. A *kilogram* is a measure of the *mass*, or amount of matter, in an object.

Correction/Completion

2. Answer:
   a. No matter where you go in the universe, your *mass* measured in *kilograms* will be the same.
   b. On Earth, a mass of 50 kg weighs approximately *500 N*.

Short Response

3. Answer:
   a. A spring scale
   b. An equal-arm balance
   c. An equal-arm balance

Illustrative

4. Answer:

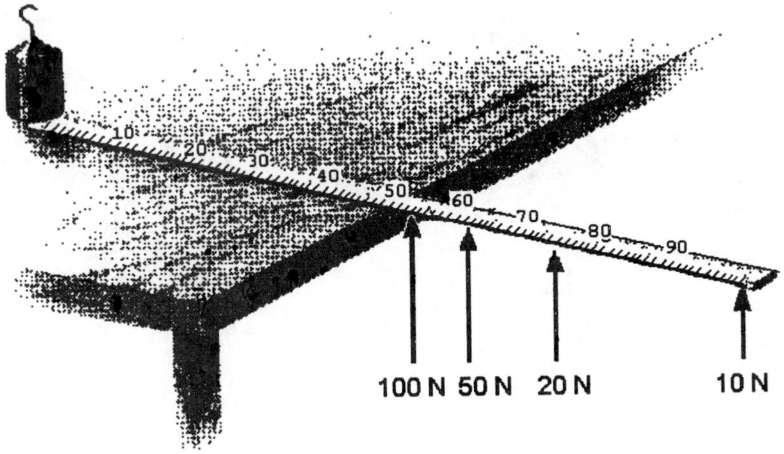

Numerical Problem

5. Answer:
   a. Each marble has a mass of about 10 g.
   b. She can conclude that the object weighs about 5 N.

   Difficulty: 3

Short Response

6. Answer:
   Sample answer: Cut the string so that it is the exact length of one side of the cube. Then fold the string in half, cut it, and lay one half along one side of the cube. Now slice the cube into two equal halves, using the balance to ensure they are equal. The resulting halves should each have a mass of 500 g. If you repeat the process of dividing and balancing again, you will have two 250 g masses. Repeat the process once more with a 250 g mass, and the result is a 125 g slice of cheese.

   Difficulty: 4

Chapter 12 Assessment

Word Usage

1. Use all of the following terms in one or two sentences to show how they are related: motion, direction, and frictional force.

Short Response

2. Suppose that you and your family are traveling by car on a mountain highway when you encounter icy and snowy patches on the road.
   a. Using what you've learned about friction, explain why the driving conditions are dangerous.

   b. Offer some ideas on how to combat the slippery roads.

3. Juan conducted an experiment with a baseball, a ramp, and three different surfaces. He rolled the ball 3 times on each surface and graphed his results as shown below.
   a. What does the graph show?

____

____

____

____

____

____

____

____

____

____

   b. The following table contains the data that Juan collected each time he rolled the ball on a surface. Use the graph to fill in the missing numbers.

Surface	Distance traveled by the ball (m)		
	Trial 1	Trial 2	Trial 3
grass	2.0		2.1
artificial turf		3.8	3.9
hard dirt	3.5	3.6	

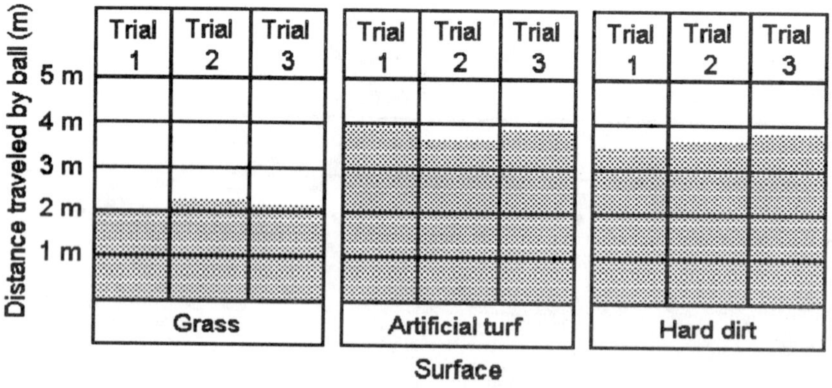

Illustrative

4. Joe wants to build a fire, but he forgot to bring matches with him on his camping trip. Shown below are the items he did bring with him. Can you solve Joe's dilemma by suggesting how he could use these items to build a fire? Explain your solution.

Short Response

5. Motor oil is very important for cars and trucks. Without motor oil, engines would quickly break down. Explain how motor oil protects car and truck engines.

Answers to Chapter 12 Assessment

Word Usage

1. Answer: Sample answer: The *direction* of a *frictional force* opposes the *direction* of *motion*

Short Response

2. Answer:
   a. Friction allows the wheels of a car to move because the ground pushes in the opposite direction of the tires. Ice and snow make the road smooth and reduce friction, making it difficult for a car's tires to turn normally. This is why the road seems slippery and dangerous.

   b. Possible methods for increasing the friction between the road and the tires include using snow tires or chains or waiting until road-maintenance crews add sand to the road.

Graphic

3. Answer:
   a. The graph shows that friction varies with different kinds of surfaces. On the grass, the roughest surface, friction was the greatest, so the ball rolled only 2.1 m. On the artificial turf, the smoothest surface, friction was the least, so the ball rolled the farthest, 3.9 m. The hard dirt created less friction than the grass but more friction than the artificial turf, and the average distance traveled by the ball was 3.6 m.

   b.

	Distance traveled by ball (m)		
Surface	Trial 1	Trial 2	Trial 3
grass	2.0	**2.2**	2.1
artificial turf	**4.0**	3.8	3.9
hard dirt	3.5	3.6	**3.7**

Illustrative

4. Answer:
   Possible answers: Find a flint stone and hit it sharply against the steel hammer or hatchet. Twist a tent stake in a hole in a flat wooden surface (cut by the hatchet from a tree) to generate enough friction to light a fire. Use the magnifying glass, the sun, and some tinder (such as the rope) to create enough heat energy to start a fire, etc.

   Difficulty: 3

Short Response

5. Answer:
Motor oil coats the moving metal parts that rub against each other with a thin film. Without the thin layer of oil separating them, the engine parts would touch each other. The resulting friction would cause a tremendous buildup of heat, and the parts would melt or break.

Difficulty: 4

Chapter 13 Assessment

Correction/Completion

1. The following statements are incorrect or incomplete. Your challenge is to make them correct and complete.
   a. Inertia is needed to start an object in motion.

   ____

   ____

   b. If forces acting on an object at rest are balanced, the object tends to move in a straight line.

   ____

   ____

Short Response

2. Imagine that you are riding in a taxicab and wearing your seat belt. Show your understanding of inertia and the effects of force by describing what happens to you during the following situations.
   a. The taxi driver applies the brakes, and the car comes to a sudden stop.

   b. The driver puts a foot to the gas pedal, and the car gains speed rapidly.

   c. The car makes a sharp turn to the left without slowing down.

   d. Having come out of the turn, the car proceeds along a straight, smooth, level stretch of road at a constant speed.

Graphic

3. The graph below represents part of a bicycle ride that Kim took.

a. At what point did Kim suddenly see an object in the road? Why do you think so?

____

____

b. At what point was Kim pedaling steadily on a flat surface? How can you tell?

____

____

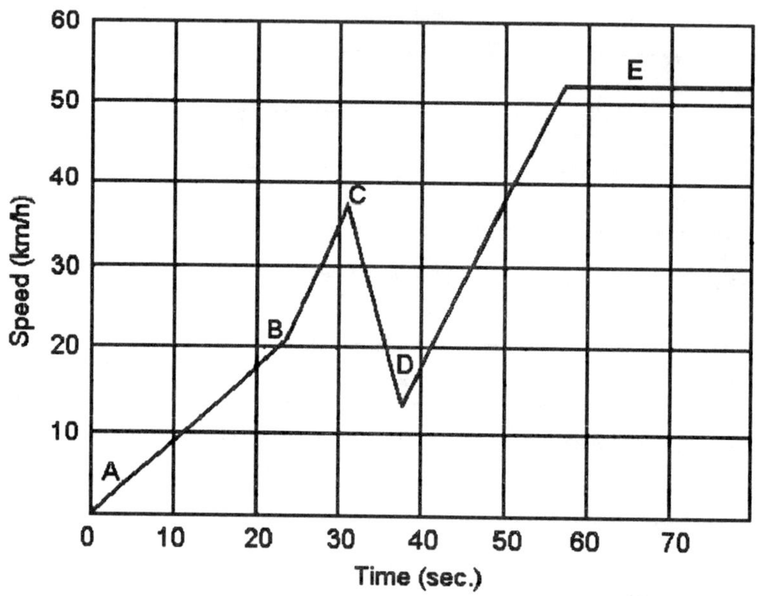

Data for Interpretation

4. The table below describes a runner in a 100 m race. From the information given, answer the questions that follow.

Time (s)	Distance from the start (m)	Time (s)	Distance from the start (m)
0	0	7	64
1	4	8	76
2	10	9	88
3	18	10	100
4	28	11	106
5	40	12	110
6	52	13	110

a. At which times were the forces on the runner balanced? How can you tell?

____

____

____

____

____

b. At which times were the forward forces increasing, indicating that the runner changed speed?

____

____

____

____

____

Illustrative

5. This illustration shows one skier at several points. The skier's speed is given at each of these points. Describe the balance of forces acting on the skier at each of the different points along the skier's journey.

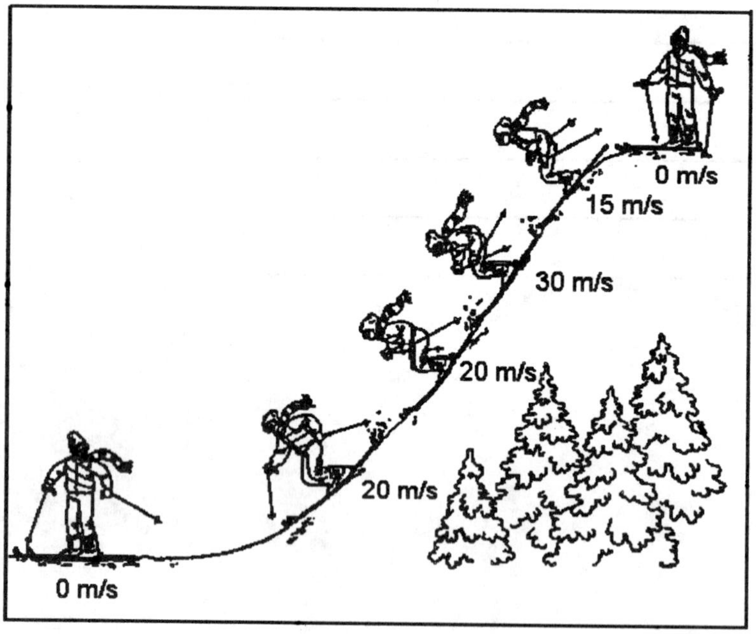

Answers to Chapter 13 Assessment

Correction/Completion

1. Answer:
   a. Sample answer: *A force* is needed to start an object in motion.
   b. Sample answer: If forces acting on an object at rest are balanced, the object *will remain at rest.*

Short Response

2. Answer:
   a. Because of your inertia, you will continue to move forward and press against the seat belt. It will feel like the seat belt has tightened.
   b. Because of your inertia, you will tend to remain at rest and press against the back of your seat.
   c. Your inertia will tend to keep you moving straight relative to the ground while the car changes direction relative to the ground. Therefore, you will move slightly toward the right side of the car, continuing your original forward motion.
   d. You will not experience any unbalanced forces.

Graphic

3. Answer:
   a. At point *C*, because that is where the graph shows the beginning of a sharp decline. This decline likely indicates the fact that Kim suddenly applied the brakes and slowed down sharply.
   b. At point *E* there is a straight line representing a constant speed of 52 km/h, so Kim must have been maintaining a fast, steady speed on a flat surface.

Data for Interpretation

4. Answer:
   a. 0 s; from 4 s to 10 s; from 12 s to 13 s. At 0 s and from 12 s to 13 s the runner was at rest. During each of the intervals from 4 s to 10 s, the runner traveled 12 m, indicating a constant speed. Therefore, the forces on the runner were balanced.
   b. During each interval from 1 s to 4 s, the distance traveled by the runner increases. For example, the distance traveled from 1 s to 2 s was 6 m, and the distance traveled from 2 s to 3 s was 8 m, indicating an increasing speed.

Illustrative

5. Answer:
   0 m/s: balanced forces; skier at rest
   0 m/s-15 m/s: unbalanced forces; skier accelerating
   (gravity > friction)
   15 m/s-30 m/s: unbalanced forces; skier accelerating
   (gravity > friction)
   30 m/s-20 m/s: unbalanced forces; skier negatively accelerating
   (friction > gravity)
   20 m/s-20 m/s: balanced forces; skier maintains constant speed
   (friction = gravity)
   20 m/s-0 m/s: unbalanced forces; skier comes to a stop
   (friction > gravity)

Unit 4 End-of-Unit Assessment

Word Usage

1. Alan and Marge spent the afternoon jumping on a trampoline. Describe some of their experiences using the following terms: gravitational force, frictional force, and elastic force.

____

____

____

____

____

2. Describe any situation of your choice that shows how forces occur in pairs. Use the words *action* and *reaction*. For example, you might describe the interaction between a tennis racket and a tennis ball as it is being hit.

____

____

____

____

Correction/Completion

3. Make the statement below correct and complete. There may be more than one way to do this.

   Inertia is the tendency of an object in motion to

   ____

Short Response

4. Give two examples in which friction is helpful.

____

____

____

5. Give two examples in which friction is harmful.

Illustrative

6. Anthony gives Laura a shove, as shown below. Use the space provided to describe what happens. Who moves faster? Why?

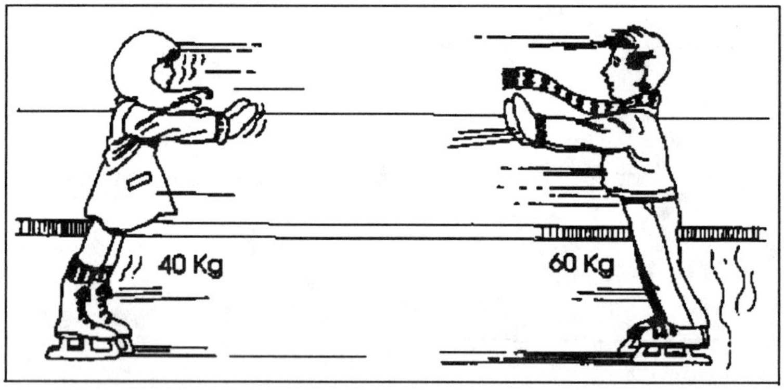

7. a. Circle the places where friction is present in the picture below.
   b. How might you decrease friction to make the job easier?

____

____

____

Graphic

8. The graph shows the stretch of a spring when it is pulled by different forces. Use the graph to answer the questions that follow.

   a. What is the stretch when no force is applied? __________

   b. When the stretch is 20 mm, how great is the force? __________

   c. What force and what stretch does point *P* on the graph represent?

   ____

   d. When the force increases by 2 N, how much does the stretch increase? ____________

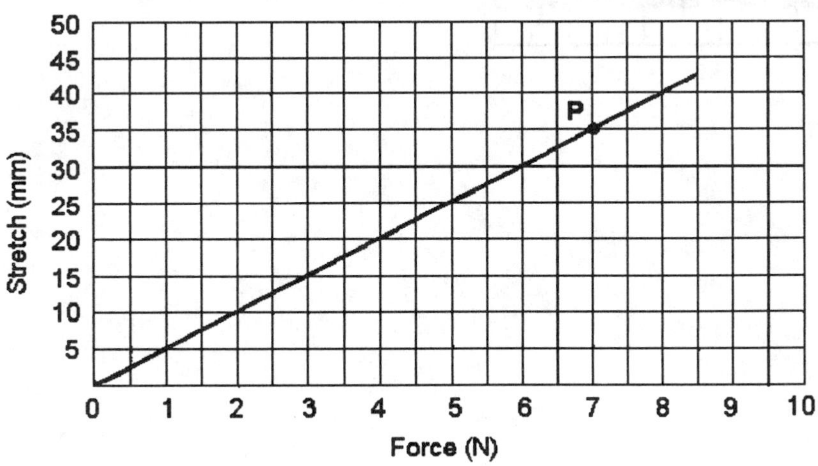

9. A car is traveling at 28 km/h. The driver applies the brakes. The graph shows how far the car travels before stopping under five different circumstances.

a. In which case was the frictional force greatest? What was the stopping distance in this case?

___

___

b. In which case was the frictional force the least? What was the stopping distance in this case?

___

___

c. To stop on ice in the shortest distance, would you use regular tires, regular tires with chains, or studded tires?

___

___

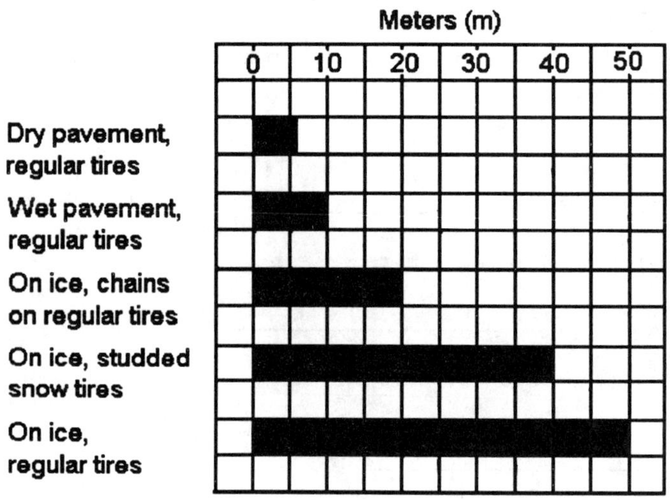

Short Essay

10. Use the diagram below to explain to a friend why the moon circles the Earth. In your explanation, mention the force that attracts the moon to the Earth. State the direction in which the moon would move if there were no force pulling it toward the Earth.

____

____

____

____

____

____

____

____

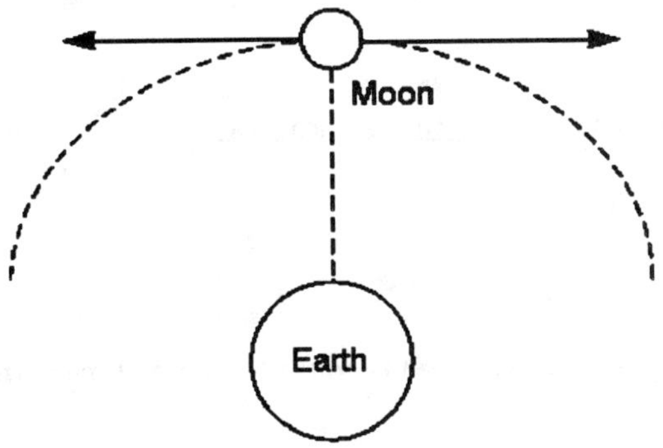

Answers to Unit 4 End-of-Unit Assessment

Word Usage

1. Answer:
   Sample answer: Each time Alan and Marge jumped, the trampoline returned to its original shape because of *elastic force*. The Earth exerted a *gravitational force* on their bodies that pulled them downward. *Frictional force* between their feet and the trampoline kept them from slipping.

2. Answer:
   Sample answer: When the racket strikes the ball, an *action* force occurs, and the ball is pushed in the direction of the swing. The *reaction* force is the force of the ball pushing the racket in the opposite direction of the swing.

Correction/Completion

3. Answer: Inertia is the tendency of an object in motion to <u>stay in motion</u>.

Short Response

4. Answer:
   Sample answer: Friction is helpful when running on a basketball court and when pulling on a rope.

5. Answer:
   Sample answer: Friction is harmful when it wears down the treads on a tire and when it produces excessive heat inside an engine.

Illustrative

6. Answer:
   Both move backward. The force of Anthony on Laura equals the force of Laura on Anthony because all forces occur in pairs. Laura moves faster than Anthony because Anthony has greater mass; it takes a greater force to overcome his inertia. (The greater the mass, the greater the force needed to move the object.)

7. Answer:
b. Sample answer: The table could be lubricated or the box could be set on rollers to decrease friction.

Graphic

8. Answer:
a. 0 mm
b. 4 N
c. 7 N; 35 mm
d. 10 mm

9. Answer:
a. Dry pavement, regular tires; about 6 m
b. On ice, regular tires; 50 m
c. Regular tires with chains

Difficulty: 3

Short Essay

10. Answer:
Sample answer: The moon is in motion, but it does not move in a straight path. Instead, it reacts to the constant gravitational force of the Earth. This places a force on the moon that pulls it toward the Earth. The force constantly interrupts the moon's tendency to travel in a straight line, keeping the moon at a constant distance from the Earth as it moves and creating a circular path. If no gravitational force existed, the moon would continue moving in a straight line, as illustrated by the arrow in the diagram.

Difficulty: 4

Unit 4 Activity Assessment

Activity Assessment

1. **Free Fall**

Teacher's Notes

Overview
Students design their own experiment to test Galileo's hypothesis that two objects with different masses fall at the same rate.

Materials
(per activity station)

- a golf ball
- a basketball
- 2 pieces of white paper
- a balance

Preparation
Prior to the assessment, equip student activity stations with the materials needed for each experiment.

Time Required
Each student should have 30 minutes at the activity station and 10 minutes to complete the Experiment Design Form.

Performance
At the end of the assessment, students should turn in the following:

- a completed Experiment Design Form

Evaluation
The following is a recommended breakdown for evaluation of this Activity Assessment:

- 25% appropriate use of materials and equipment
- 40% thorough and logical experimental design
- 15% use of control and variables
- 20% ability to form a logical conclusion based on observation

2. **Free Fall**

Here's a puzzle for you:

A dump truck and a pebble are on top of your house. Would the pebble fall to the ground faster than the dump truck? Use your sharp eyes, careful measurements, and scientific mind to draw some conclusions about the rate at which objects fall. Remember that more than one variable may be affecting your results.

Before You Begin . . .

As you work through the tasks, keep in mind that your teacher will be observing

- how you use the materials and equipment
- how clear and complete your experiment is
- how well you draw conclusions from your observations and data

Now you are ready to solve the puzzle!

Task 1 Write a hypothesis that gives a scientific answer to the question, Do objects of different masses fall at the same rate? Record your hypothesis on the Experiment Design Form.

Task 2 Using both balls and both pieces of paper, plan and carry out tests to support your hypothesis. For accurate results, you may need to perform your test more than once for each of the materials you use. Outline the steps of your experiment and record your data on the Experiment Design Form.

Task 3 Complete the Experiment Design Form. Be sure to discuss whether your tests supported your hypothesis. Also include a statement or two about the variable(s) that affect(s) free fall.

3.

Experiment Design Form

Hypothesis
Plan
Data
Conclusion

Answers to Unit 4 Activity Assessment

1. Answer: Not applicable (teacher's notes)

2. Answer: Not applicable (student's notes)

3. Answer:

Experiment Design Form

Hypothesis Sample answer: Two objects of different masses will fall at the same rate.
Plan Sample plan: 1. Find the masses of the golf ball, the basketball, and the pieces of paper. 2. Hold the golf ball and the basketball at arm's length. Drop both at the same time. See if they land together or if one hits the floor before the other. Record your observations. Repeat three times. 3. Crumple one piece of paper. Repeat step 2 using the crumpled paper and the basketball. Record all results. Next, use the crumpled paper and the golf ball, and try the experiment again. Record all observations. 4. Using the crumpled paper and the flat paper, repeat the test again. Record your results. Then test the flat paper against both balls. Record all results.
Data Data will vary depending on the hypothesis and plan, but ensure that a control is used and that all results are clear and complete.
Conclusion Sample conclusion: When the shape of the object is similar, objects of different masses will fall at the same rate. However, the golf ball, basketball, and crumpled paper fell faster than the piece of flat paper. So the shape of an object can affect the rate of fall if variables other than the force of gravity (such as air resistance) are acting on that object.

Unit 4 SourceBook Assessment

Multiple Choice

1. The car was traveling at 88 km/h. This is an example of
   a. speed. b. velocity. c. acceleration. d. reference point.

2. The car was traveling south at 88 km/h. This is an example of
   a. speed. b. velocity. c. acceleration. d. reference point.

3. The car was traveling south at 88 km/h away from the city. In this example, the city indicates the
   a. field of view. b. direction. c. location. d. frame of reference.

4. If a cyclist rides up a hill during a bike tour, his or her speed decreases. In scientific terms this is called
   a. acceleration. b. tiring. c. a change in velocity. d. a and c only.

5. If Bill were riding in a car moving 88 km/h, and if a bee inside the car were flying forward at 3 km/h, how fast would Bill think the bee was moving?
   a. 88 km/h   b. 3 km/h   c. 91 km/h

6. Bill was riding in a car moving 88 km/h, and a bee inside the car was flying forward at 3 km/h. Using the Earth as the frame of reference, how fast was the bee moving?
   a. 88 km/h   b. 3 km/h   c. 91 km/h

7. If friction did not exist and you rolled a ball on the floor, which of the following would the ball do?
   a. roll farther than it would if friction existed, and then stop
   b. stop immediately
   c. roll at the same velocity until it met an obstacle
   d. gradually speed up until it reached the speed of light

8. Swimmers often shave their bodies before a big race so that they swim faster. Why does this help them swim faster?
   a. it increases velocity
   b. it decreases friction
   c. it changes their frame of reference
   d. a and b only

9. Which would hit the ground sooner, a ball dropped from a 90 m cliff or a ball rolled off the same cliff at 30 m/s?
   a. the dropped ball   b. the rolled ball
   c. They would each take the same amount of time to hit the ground.

10. If you are the passenger in the front seat of a car and the driver drives in very tight, fast circles to the left, your body would automatically lean
a. backward against your seat.
b. to the left.
c. to the right.
d. You would not be affected by the car's movement.

11. You are standing in your yard and you throw a ball. It follows the path shown. The curved line is called the ball's
a. centripetal force. b. reference line. c. trajectory. d. projectile.

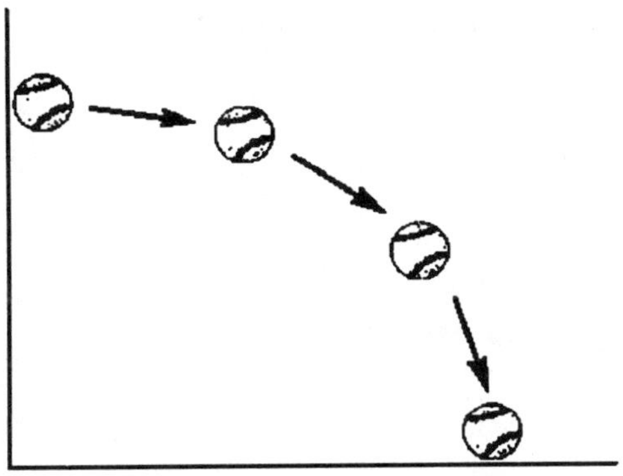

12. In the illustration above, what would the ball be called?
a. circular motion b. projectile c. trajectory d. reference point

13. Suppose you have a red ball and a blue ball. If you held a ball in each hand and dropped the red one straight down from shoulder height and threw the blue ball 60 cm in front of you at shoulder height, which ball would reach the ground first?
a. the red ball
b. the blue ball
c. Not enough information is provided.
d. They would reach the ground at the same time.

14. The football sits on the tee waiting for the kicker to kick it. Unless the wind knocks it off, the ball will sit there all night until the kicker's foot hits the ball. This best describes the property of
a. inertia. b. equal motion. c. velocity. d. friction.

15. If force remains the same and mass increases, then acceleration
a. stays the same. b. increases. c. decreases.

16. If mass stays the same and acceleration increases, then force must have
a. stayed the same. b. increased. c. decreased.

17. A basketball player runs down the court and jumps up to the hoop. As soon as he leaves the floor, gravity begins to pull him back down, yet his __________ keeps him moving upward to slam-dunk the ball for two points.
a. mass b. velocity c. speed d. momentum

18. What law of nature explains why the planets in our solar system continue to orbit the sun?
a. inertia b. acceleration c. gravity d. momentum

19. If you stood on your bathroom scale on the surface of the moon, the scale would show that you weighed (more than/less than/the same as) you do on Earth. (Circle the correct answer.)

20. On the moon, you would have (more/less/the same) mass as you do here on Earth. (Circle the correct answer.)

True/False

21. If a racehorse runs around the last bend of the racetrack without changing its speed, the horse is considered to be accelerating.
a. true b. false

22. Your clothes undergo centripetal force while spinning in a washing machine.
a. true b. false

23. For every action, there is an equal and opposite reaction.
a. true b. false

Short Response

24. If you dropped a bowling ball and a baseball from a height of 20 m, which would reach the ground first? Explain.

____

____

____

____

25. If you dropped a bowling ball and a feather from a height of 20 m, which one would reach the ground first? Explain.

____

____

____

____

____

Answers to Unit 4 SourceBook Assessment

Multiple Choice

1. Answer: a. speed.
2. Answer: b. velocity.
3. Answer: d. frame of reference.
4. Answer: d. a and c only.
5. Answer: b. 3 km/h
6. Answer: c. 91 km/h
7. Answer: c. roll at the same velocity until it met an obstacle
8. Answer: b. it decreases friction
9. Answer: c. They would each take the same amount of time to hit the ground.
10. Answer: c. to the right.
11. Answer: c. trajectory.
12. Answer: b. projectile
13. Answer: d. They would reach the ground at the same time.
14. Answer: a. inertia.
15. Answer: c. decreases.
16. Answer: b. increased.
17. Answer: d. momentum
18. Answer: c. gravity
19. Answer: <u>less than</u>
20. Answer: <u>the same</u>

True/False

21. Answer: a. true

22. Answer: a. true

23. Answer: a. true

Short Response

24. Answer:
    They would hit at approximately the same time. According to Newton's law of gravity, everything on Earth falls at the same rate (not taking into account air resistance).

25. Answer:
    The bowling ball would hit first because the feather would have more air resistance working against it. In this case, because the shapes of the objects differ greatly, air resistance cannot be ignored; the feather would float to the ground more slowly.

Unit 4 Extra Assessment Items

Word Usage

1. Describe a batter hitting a ball. In doing so, use all of the following words: force, agent, receiver, and effect.

2. Explain what happens to a moving bicycle and a rider when the forces on them are balanced and when they are unbalanced. Use the following words: balanced forces, unbalanced forces, and rolling friction.

3. Use these terms to answer the questions that follow.

decrease	equal-arm balance	agent	kilogram
newton	diving board	balanced force	matter
ball bearings	buoyant	unbalanced force	push
pull	decrease	noncontact	contact
elastic force	effect	sliding friction	lubricate
motion	gravitational force	magnetic force	boat
Newton	spring scale	receiver	

a. Choose two terms from the list to use with the word *force*.

1. ______________

2. ______________

3. ____force____

Write one or two sentences using these three terms to show that you know how the terms are related.

____

____

____

b. Choose two terms from the list to use with the word *weight*.

1. ______________

2. ______________

3. ____weight____

Write one or two sentences using these three terms to show that you know how the terms are related.

____

____

____

c. Choose two terms from the list to use with the word *friction*.

1. ______________

2. ______________

3. ____friction____

Write one or two sentences using these three terms to show that you know how the terms are related.

____

____

____

Correction/Completion

Make the statements below correct and complete. There may be more than one way to do this.

4. The weight of a 1000 g brick is 1 N.

5. An object in motion with no unbalanced forces acting on it will eventually come to a stop.

6. Inertia is based on an object's weight; the greater the weight, the less inertia the object has.

7. The gravitational force of one object on another decreases as the masses of the objects increase. However, the gravitational force increases when the two objects move apart.

8. A balanced force acting on an object in motion can cause it to change its speed or direction.

9. a. Complete the graph below to show the stretch of a spring for forces up to 700 N.

b. What is the force when the stretch is 65 mm? ________

c. When the force is 250 N, how great is the stretch? ________

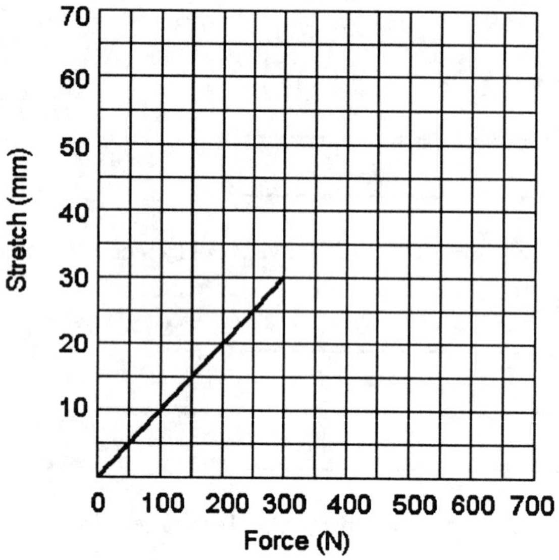

Short Essay

10. How could you use a skateboard, a force meter, and a piece of string to show that rolling friction is less than sliding friction?

____

____

____

____

____

____

____

____

11. Imagine that you are pushing a grocery cart. Write a story about this activity that describes the forces acting on the cart, as well as the effects of balanced and unbalanced forces on the cart.

Short Response

12. Name the following kinds of forces:
    a. the force between two masses
    b. the force between a comb (after combing your hair) and small bits of paper
    c. the force applied by bicycle brakes
    d. the force that supports you on a bathroom scale
    e. the force that supports a boat on water

13. Provide an example for each of the following:
a. a noncontact force

b. an elastic force

c. a balanced force

d. inertia

14. What device should you use to measure the following:
a. the weight of a brick on Earth

b. the mass of a brick in space

c. the gravitational force pulling on a table

15. Suppose that you are on a camping trip and find yourself needing to move a heavy wooden box along rough ground. What are two ways you might devise to reduce the friction that is preventing you from moving the box? (Be realistic! If you use any extra materials, they should be available in your camping area.)

a. ______________________________

____

____

____

____

b. ______________________________

____

____

____

____

16. Which would fall to the ground more slowly—an inflated balloon or an uninflated balloon? Explain.

____

____

____

____

____

17. Describe the effects that a force can have on an object.

____

____

____

____

____

18. Describe the difference in the effects caused by balanced and unbalanced forces on an object.

Illustrative

19. Illustrate the story you wrote about pushing the grocery cart, using arrows to represent the forces.

20. In the following illustration, identify the agent of the force and the receiver of the force, and predict or describe the effect of the force on the receiver.

agent:

receiver:

predicted effect on receiver:

21. In the following illustration, identify the agent of the force and the receiver of the force, and predict or describe the effect of the force on the receiver.

agent:

receiver:

predicted effect on receiver:

22. In the following illustration, identify the agent of the force and the receiver of the force, and predict or describe the effect of the force on the receiver.

agent: ______ ______ receiver: ______ ______ predicted effect on receiver: ______ ______ ______ ______	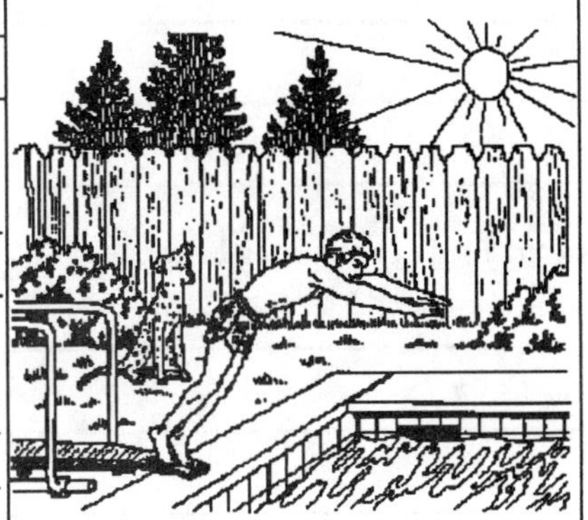

23. In the following illustration, identify the agent of the force and the receiver of the force, and predict or describe the effect of the force on the receiver.

What happens when the ball contacts the bat? agent: ______ ______ receiver: ______ ______ predicted effect on receiver: ______ ______ ______	

24. Did any of the examples in the previous questions illustrate a noncontact force? If so, identify the illustration and the noncontact force illustrated.

______

______

25. Use the illustration on the next page to answer the questions that follow.

a. List at least one example of gravitational force.

____

____

____

____

b. List at least one example of frictional force.

____

____

____

____

d. Identify an example of buoyant force.

____

____

____

____

e. Identify an example of elastic force.

____

____

____

____

f. Identify an example of a contact force.

____

____

____

____

g. Identify an example of a noncontact force.

____

____

____

____

h. Find an example of how inertia resists the effects of another force.

i. Find an example where friction has been reduced.

j. Describe the interaction of two of the objects shown in terms of action and reaction forces.

26. For the following illustration, answer the questions.

<table>
<tr><td>The parachutist is falling at a constant speed. What are the opposing forces?
____
____

Are the forces balanced? If not, which is greater?
____
____
____
____
____
____
______________________________</td><td></td></tr>
</table>

27. For the following illustration, answer the questions.

<table>
<tr><td>What are the opposing forces?
____
____

Are the forces balanced? If not, which is greater?
____
____
____
____
____
____
______________________________</td><td></td></tr>
</table>

28. The illustration shows a 10 kg block resting on a table. Use this information to answer the following questions.

a. What causes the downward force?

____

b. What provides the upward force on the block?

____

c. Is the illustration labeled correctly? Explain.

____

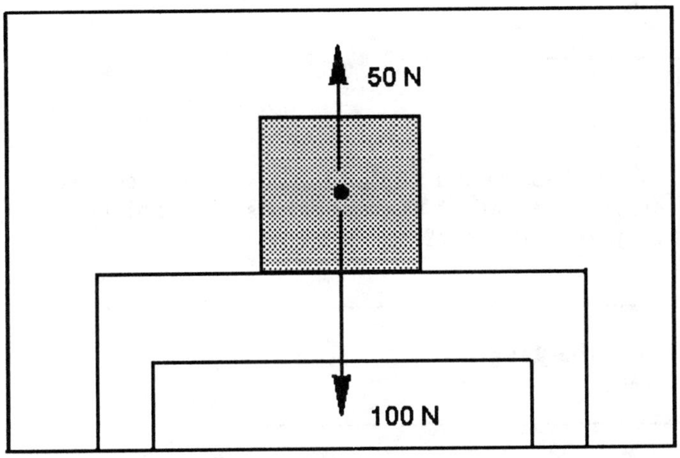

29. Correct the figure below so that it supports the following statement:

The 10 kg block is being pulled steadily to the right along the table. There is no acceleration (increase in speed).

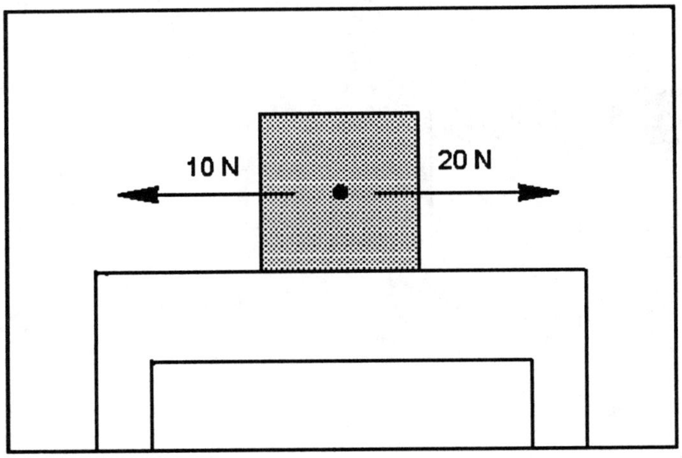

30. Correct the figure below so that it supports the following statement.
The 10 kg block is speeding up and moving to the right along the table.

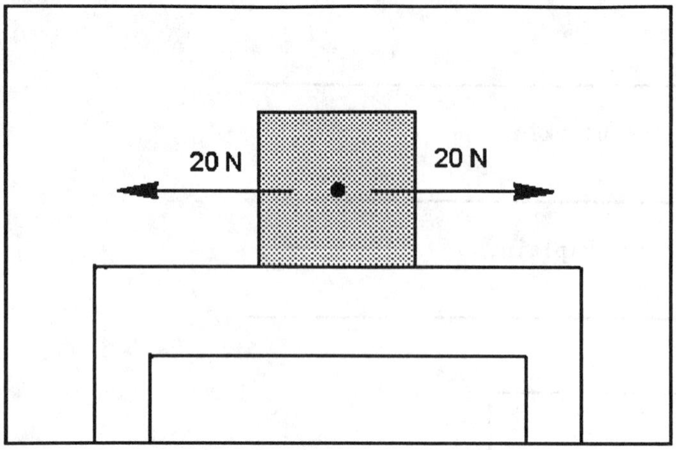

Graphic

31. A ball was rolled across three different surfaces, and the distances it rolled were recorded. Consider the graph of the results and answer the questions that follow.
a. On which surface was friction probably the greatest?

____

b. On which surface was friction probably the least?

____

c. What variables had to be controlled to make this a fair test?

____

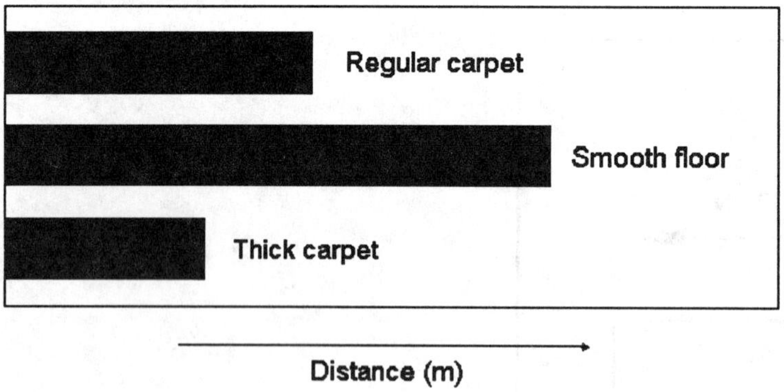

Data for Interpretation

32. Use the chart below to answer the questions that follow.

Stretch (mm)	Force (N)
0	0
10	12
20	24
30	36
40	48
50	60

a. What is the force required to stretch the spring 30 mm? ______

b. If the stretch were 35 mm, what would the force be? ______

c. How far would a force of 30 N stretch the spring? ______

Numerical Problem

33. Use what you've learned about force and weight to answer the following questions.
a. What is the weight of a 12 kg box on Earth? Show your work.

____

b. What is the weight of the box on the moon (where the gravitational force is one-sixth of that on Earth)? Show your work.

____

34. Aleisha and Ubaldo are racing their planes in a statewide competition. Ubaldo's plane has a top speed of 750 km/h and can fly 1200 km before running out of fuel. Aleisha's plane has a normal top speed of 675 km/h and a range of 1100 km. For this race, however, Aleisha has created a new, streamlined design for her plane in order to reduce air friction. In doing so, she has improved her top speed by 12 percent and her fuel efficiency by 8 percent.

a. Can Aleisha's new plane fly faster than Ubaldo's plane? Show your work.

b. Can Aleisha's plane fly farther than Ubaldo's plane?

c. If the race is 1200 km long, can Aleisha finish without stopping to refuel?

Performance Task

35. Complete the following activity:

You Will Need

- a spring scale
- a glass
- a chair
- a bathroom scale calibrated in newtons
- a book
- a washer

What to Do

a. Estimate the weight (in newtons) of each object below, and record your estimate in the table.

b. Measure the weight of each object and record it in the table.

c. Calculate the mass of each object and record it in the table.

Object	Estimated weight (N)	Measured weight (N)	Calculated mass (kg)
a book			
a glass			
you			
a washer			
a chair			

d. If you were to take these objects with you on a space mission, what would happen to the masses of these objects as you left the Earth's surface?

____

____

____

e. What would happen to the weights of these objects?

____

____

____

36. Complete the following activity:

You Will Need

- a balloon
- a rubber band
- a spring scale calibrated in newtons
- other materials you select

What to Do

a. Using the balloon, construct a simple force meter. Calibrate it with the spring scale. Plot a graph that shows how much the rubber band stretches when different amounts of force are used. Remember to label the *x*- and *y*-axes.

b. Use your force meter to measure small forces in newtons. Report the results of your measurements in the following table.

Agent	Receiver	Type of Force	Amount of force (N)

Answers to Unit 4 Extra Assessment Items

Word Usage

1. Answer:
   Sample answer: When the bat strikes the ball, it exerts a pushing *force*. The bat is the *agent* of the force, and the ball is the *receiver* of the force. The *effect* of the force is a change in the direction and speed of the ball.

2. Answer:
   Sample answer: *Balanced forces* that are exerted on a moving bicycle and rider keep the bicycle moving in steady motion in a straight line. *Rolling friction* occurs between the ground and the bicycle's tires. If *unbalanced forces* are exerted, the bicycle may swerve, stop, slow down, or speed up.

3. Answer:
   a. A wide number of correct answers are possible. Make sure that the response is clear and logical.
   b. A wide number of correct answers are possible. Make sure that the response is clear and logical.
   c. A wide number of correct answers are possible. Make sure that the response is clear and logical.

Correction/Completion

4. Answer: The weight of a 100 g brick is 1 N; or the weight of a 1000 g brick is 10 N.

5. Answer: An object in motion with no unbalanced forces acting on it will *continue in motion*.

6. Answer:
   Inertia is based on an object's *mass*; the greater the *mass*, the *more* inertia the object has.

7. Answer:
   Sample answer: The gravitational force of one object on another *increases* as the masses of the objects increase. However, the gravitational force *decreases* when the two objects move apart.

8. Answer:
   An *unbalanced* force acting on an object in motion can cause it to change its speed or direction.

9. Answer:
   a. Because the amount the spring stretches is directly proportional to the force added to it, graphs should indicate a straight line.
   b. 650 N
   c. 25 mm

Short Essay

10. Answer:
Sample answer: Attach one end of a string to a skateboard and the other end to a force meter. Measure the force needed to keep the skateboard rolling. Then turn the skateboard over and measure the force needed to drag the skateboard on its back. Compare the forces you measured.

11. Answer:
Answers will vary, but students should indicate the roles of friction and of the human's pushing the cart in this scenario. Students should also demonstrate the understanding that unbalanced forces lead to different kinds of acceleration.

Short Response

12. Answer:
a. Gravitational force
b. Electrical force
c. Frictional force
d. Elastic force (of the spring in the scale)
e. Buoyant force

13. Answer:
a. Sample answer: The force (magnetic force) that attracts paper clips to a magnet
b. Sample answer: The force that causes a spring to return to its original, relaxed length after it has been stretched or compressed
c. Sample answer: An ice skater moving in a straight path at a constant speed
d. Sample answer: A passenger continuing to move forward in a car that is stopping

14. Answer:
a. A spring scale
b. An equal-arm balance
c. A spring scale

15. Answer:
a. Accept all reasonable responses. Sample answer: Rolling friction is less than sliding friction. Therefore, using smooth logs to roll the heavy box would be one helpful approach. Perhaps a cart could be constructed with wooden wheels.

b. Accept all reasonable responses. Sample answers: Another approach would be to reduce the friction by smoothing the ground or the box. Still another approach might be to place smoother materials such as leaves and grass on the ground.

16. Answer:
Sample answer: The inflated balloon would fall more slowly because of air resistance. The inflated balloon has more surface area for the air to push against.

17. Answer:
Sample answer: A force acting on an object can cause the object to speed up, slow down, or change direction. If that force is balanced by an equal and opposite force, the object will either remain at rest or continue moving at a constant velocity. If the answer given includes any part of this explanation, consider it acceptable. If it includes all of these ideas, give the student a bonus.

18. Answer:
Sample answer: When the forces acting on an object are not completely balanced, the object either speeds up, slows down, or changes its direction. Otherwise, it remains at rest or continues to move at a constant speed in the same direction.

Illustrative

19. Answer:
Illustrations should correctly show arrows representing the direction of motion and of the balanced forces. It is important that any frictional forces are shown *opposing* the motion.

20. Answer:

agent:
Either the spring or the Earth's gravitational force

receiver:
The bar or weight, the spring, or all of the above

predicted effect on receiver:
The bar is bent down, or the weight is pulled down; the spring is stretched; the object is pulled toward the Earth.

21. Answer:

agent:
The gravitational force of the Earth

receiver:
The person

predicted effect on receiver:
The person accelerates downward.

22. Answer:

agent:
The person or the diving board

receiver:
The diving board or the person

predicted effect on receiver:
The person bends the board downward, or the board pushes the person upward.

23. Answer:

What happens when the ball contacts the bat?

agent:
The bat or the ball

receiver:
The ball or the bat

predicted effect on receiver:
The bat compresses and knocks away the ball, or the ball compresses and slows down the bat.

24. Answer:
The gravitational force of the Earth acting on the diver is an example of a noncontact force.

25. Answer:

a. Sample answer: Gravitational force acts downward on the parachutist and on all of the other objects shown in the illustration.

b. Sample answers: Frictional force exists between any of the wheels shown in the illustration and the ground, between the parachute and the air, between the boats and the water, between the man walking and the ground, and between the airplane and the air.

d. Sample answer: Buoyant force supports the boats in the water.

e. Sample answer: Elastic force is evident in the stretching of the trampoline.

f. All of the examples of frictional force, buoyant force, and elastic force listed above are examples of contact forces.

g. Gravitational force is a noncontact force that acts on all bodies shown in the illustration. For example, the Earth acts on the plane without being in direct contact with it.

h. Any object shown in the illustration is an example of this.

i. The best example is the use of wheels to reduce friction (skates, bicycle, truck).

j. Answers will vary. Sample answer: The weight of the person in contact with the trampoline is matched by an equal and opposite force exerted by the trampoline on the person.

26. Answer:

The parachutist is falling at a constant speed. What are the opposing forces?

Gravitational and frictional force

Are the forces balanced? If not, which is greater?

The parachutist is falling at a fairly constant speed at this point in his descent; therefore, the upward force of the parachute balances the gravitational force of the Earth, which is acting downward on the parachute.

27. Answer:

What are the opposing forces?

Elastic force and gravitational force

Are the forces balanced? If not, which is greater?

Since no motion is evident, the forces on the woman, the cat, and the chair appear to be balanced. The elastic force of the chair acting upward on the woman is equal to the gravitational force (her weight) pulling her down

28. Answer:
   a. Gravity
   b. The table
   c. No; the upward force should equal the downward force.

29. Answer:
   Correction to illustration: Since the forces are balanced, the opposing forces should be equal.

30. Answer:
   The force on the left must be less than 20 N, or the force on the right must be greater than 20 N. Unbalanced forces can cause an object in motion to speed up.

Graphic

31. Answer:
   a. On the thick carpet
   b. On the smooth floor
   c. The type of ball used, the mass of the ball, and the amount of force used to roll the ball

Data for Interpretation

32. Answer:
   a. 36 N
   b. 42 N
   c. 25 mm

Numerical Problem

33. Answer:
   a. 120 N; 12 kg x 1000 g ÷ 1 kg x 1 N ÷ 100 g = 120 N
   b. 20 N; 120 ÷ 6 = 20 N

34. Answer:
   a. Yes; 675 km/h x .12 = 81 km/h increase in speed; 81 + 675 = 756 km/h
   b. No; 1100 km x .08 = 88 km increase in range; 1100 km + 88 km = 1188 km
   c. No; she will have to stop for fuel because the plane can only fly 1188 km on one fuel tank.

Performance Task

35. Answer:
    d. They would stay the same.
    e. Because of changes in gravitational force, they would be less, eventually reaching zero (depending on location).

36. Answer:
    Answers will vary widely depending on the force meter that is designed and the objects that are used, but all data should be clear and concise.

Chapter 14 Assessment

Word Usage

1. Use the terms *force, deflection,* and *elastic response* in a sentence or two to describe what happens to a tree limb when the wind blows.

____

____

____

____

Short Response

2. For each situation below, tell whether the force being described is *shear, compressive,* or *tensile.*
   a. Meg stomped on an aluminum can and crushed it. ______________
   b. Callie twisted a wire coat hanger back and forth until it broke.

   ____
   c. Irv stretched the waistband of his shorts so that they would feel loose when he wore them. ______________

Graphic

3. The following data show what happened when a steel guitar string was stretched. Graph the data, and then answer the questions.

Tensile (stretching) force (N)	Stretch (mm)
0	0
200	1
400	2
600	3
800	4
1000	5
1200	The steel string broke.

a. What can you conclude about the relationship between force and stretching?

b. If the length of the steel string were shorter, would you expect different results? Explain.

c. If you do expect different results when a shorter steel string is used, predict what those results might be (roughly) on the graph. Use a different colored pen or pencil to show your prediction on the graph.

Illustrative

4. Suppose that you had a large supply of thin sheets of cardboard and a large supply of the hollow cardboard cylinders from rolls of paper towels. Draw a diagram below to show how you could use the cardboard sheets and cylinders to make a structure strong enough to stand on. You may cut, fold, or tape the cardboard, but you may not add any other materials.

Short Essay

5. Explain why a suspension bridge is probably a better choice than an arch bridge for spanning a wide river that has several deep areas.

Answers to Chapter 14 Assessment

Word Usage

1. Answer:
   Sample answer: *Deflection* occurs when the tree limb bends away from its normal position because of the pushing or pulling *force* of the wind. This bending is an *elastic response*.

Short Response

2. Answer:
   a. Compressive
   b. Shear
   c. Tensile

Graphic

3. Answer:
   a. Students' graphs should show that the stretching is proportional to the applied force. In other words, as equal increments of force are applied, the amount of additional stretching is the same.
   b. Yes; the string would stretch less if it were shorter.

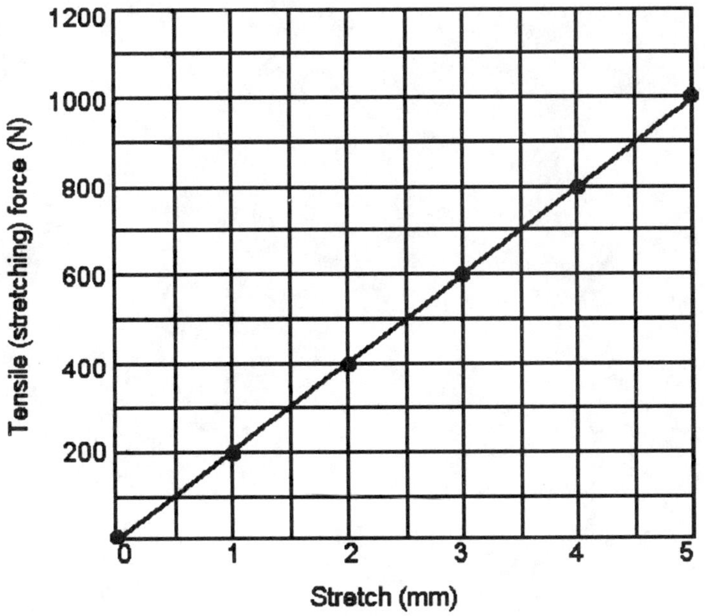

Illustrative

4. Answer:
   Answers will vary but should indicate an understanding of concepts in the chapter, such as the benefits of corrugation and the strength of tubular girders and columns.

   Difficulty: 3

Short Essay

5. Answer:
   Sample answer: A suspension bridge is much stronger over a wide span than a simple arch bridge. It is also a better choice than a bridge made of multiple arches because each arch would need two supports that touch the river bottom, which may be very deep in some spots, while the suspension bridge, which requires fewer supports that touch the bottom, can be designed so that the supports are in shallower parts of the river.

   Difficulty: 4

Chapter 15 Assessment

Word Usage

1. Use the following terms in one or two sentences to show how they are related: elevators, prestressed concrete, structural steel, and skyscrapers.

____

____

____

____

____

____

Short Response

2. For each of the architectural features listed below, name two specific buildings or general types of structures that are likely to include that feature.

   a. dome ______________________________

   ____

   b. column ______________________________

   ____

   c. arch ______________________________

   ____

3. List three factors that city planners should consider when they make plans to redevelop an old neighborhood.

____

____

____

____

____

Illustrative

4. Examine the illustration, and then answer the questions.

   a. What is this an illustration of? ______________________

   b. Describe this style of architecture.

   ____

   ____

   ____

   ____

   c. The illustration shows clear examples of two of the architectural features that you learned about in this chapter. What are they?

   ____

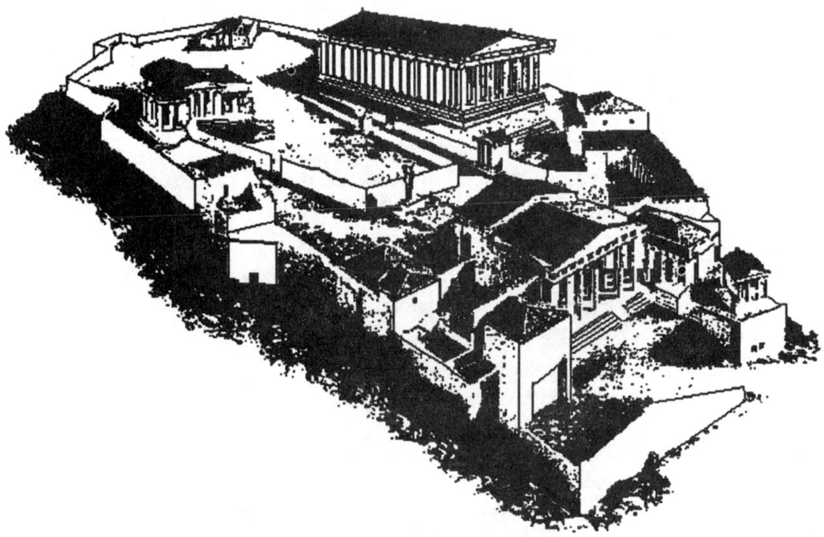

Short Essay

5. Explain how major changes in history, such as the development of farming and the growth of cities, have influenced architecture.

Illustrative

6. Frank Lloyd Wright believed in an organic architecture that links structures with nature. Walter Gropius, on the other hand, believed that architects should be able to adapt their designs to mass production by using modern materials. Choose the architectural style that you prefer, and then sketch an illustration of a house using that style. Label or describe the materials you would use to build that house.

Answers to Chapter 15 Assessment

Word Usage

1. Answer:
   Sample answer: Advances in technology, including the invention of the *elevator*, and the development of stronger building materials, such as *structural steel* and *prestressed concrete*, have made it possible to build *skyscrapers*.

Short Response

2. Answer:
   Sample answers:
   a. Hagia Sophia and some other churches, domed stadiums, igloos
   b. The Acropolis, the White House, many banks, southern mansions
   c. Roman aqueducts, the Gateway Arch in St. Louis, the Colosseum

3. Answer:
   Answers will vary but may include preservation of historic landmarks, environmental impact, pleasing design, and sturdy construction.

Illustrative

4. Answer:
   a. The Acropolis in Athens, Greece
   b. Ancient Greek architecture is characterized by scrollwork and other designs, an overall sense of balance and symmetry, and parallel columns that support the roof.
   c. Simple beams and columns

Short Essay

5. Answer:
   Sample answer: Before people knew how to grow crops, most societies were nomadic. They wandered from place to place to hunt. For example, the Native Americans followed buffalo herds; therefore, they needed simple and often portable housing, such as the tepee. When people learned to farm, they stayed in one place and built larger, sturdier, and more permanent houses out of materials such as wood and stone. As cities developed, people erected even larger and more elaborate structures such as the cathedrals and palaces of Europe.

   Difficulty: 3

Illustrative

6. Answer:
   Illustrations will vary but should be readily identifiable as one style or the other. If a student is illustrating organic architecture (Wright), he or she should show the house within the context of its natural surroundings. The materials used would most likely be wood, stone, or other materials commonly seen in nature. If a student is illustrating the International Style of architecture (Gropius), the natural environment would not be shown at all or would be portrayed as secondary to the human-built structure. Steel, glass, aluminum, and other "modern" materials will probably stand out and look very stark.

   Difficulty: 4

Unit 5 End-of-Unit Assessment

Word Usage

1. Use the terms *collapse*, *shear force*, and *structure* in a sentence or two about the Tacoma Narrows Bridge.

2. Show your understanding of the term *postmodernist* by using it in a sentence with the words *features*, *function*, and *practical*.

Short Response

3. Of the three shapes below, circle the one that can support the greatest mass.

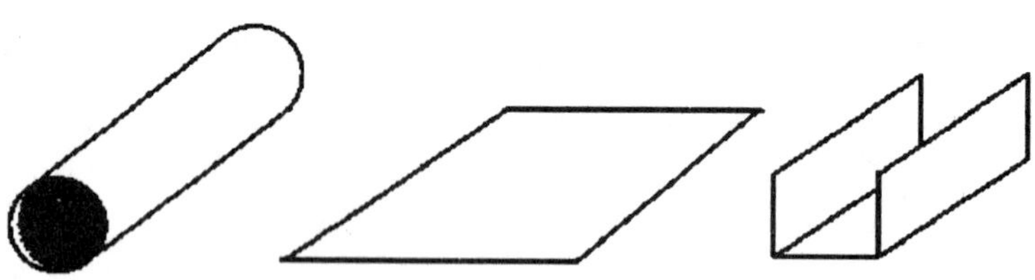

4. What is a truss?

Illustrative

5. Classify each of the following as beams or cantilevers:

   a. diving board __________

   b. box spring and mattress __________

   c. wall lamp __________

   d. tabletop __________

a.

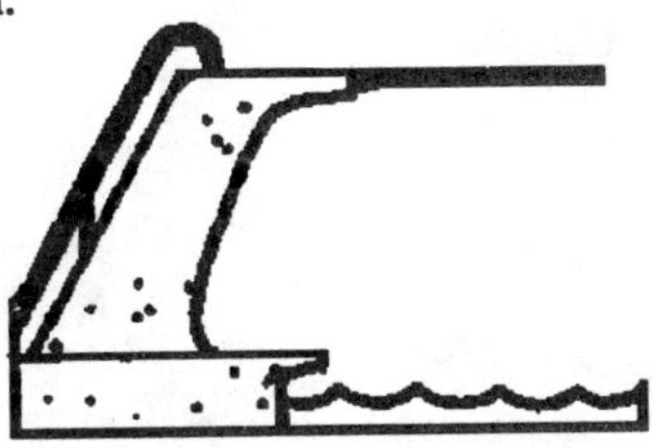

b.

c.

d.

6. Place the following labels on the correct illustrations shown below: Nomadic architecture, Ancient Greek architecture, Renaissance architecture, and Modern architecture.

a. ______________________________

b. ______________________________

c. ______________________________

d. ______________________________

a.

b.

c.

d.

Graphic

7. Monica built a model bridge using simple beams of cardboard. She added masses to the bridge and recorded the sag in centimeters. A graph of her results is shown.

   a. What was the sag when a force of 6 N was applied? __________
   b. What kind of force was acting on the bridge? ______________
   c. What do you think happened at Point A? ____________________
   d. Name three things that Monica could do to increase the strength of her bridge.

   ____

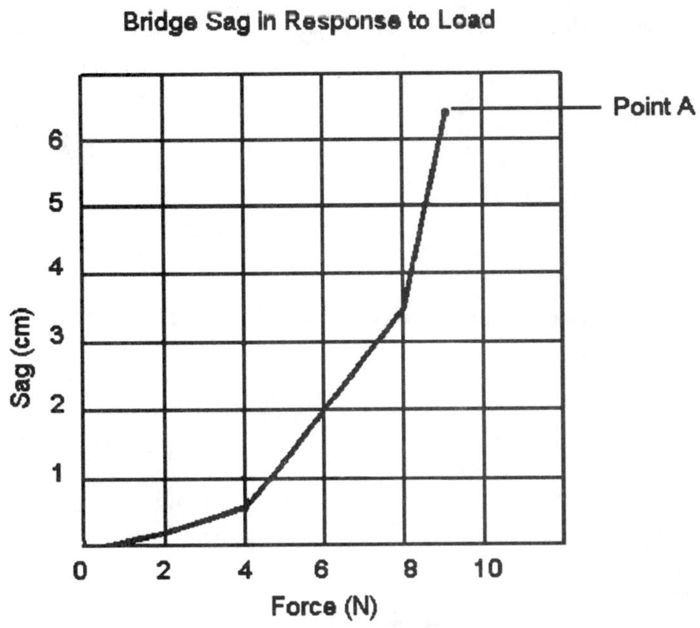

8. The graph below shows the relative strength of three structures: a simple beam, an I-beam, and an arch. In the space provided, label the graph with the correct terms.

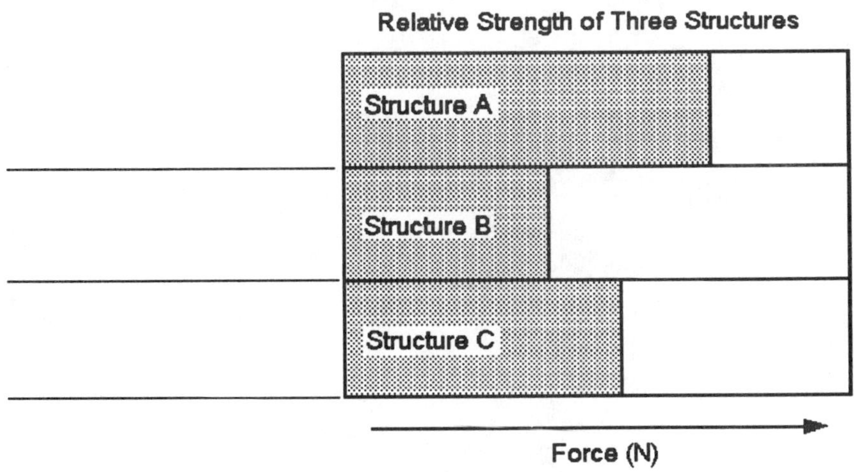

Illustrative

9. a. Draw a column affected by compressive force. Use arrows to show the direction of force.

   b. Draw a rope affected by tensile force. Use arrows to show the direction of force.

   c. Draw a roll of clay affected by shear force. Use arrows to show the direction of force.

Short Response

10. Refer to the illustration that follows when answering these questions:
    a. What are some structural features of the human body that help humans to survive?

    b. Compare the components of human-built structures (such as arches) that you learned about in this unit to specific parts of the human body.

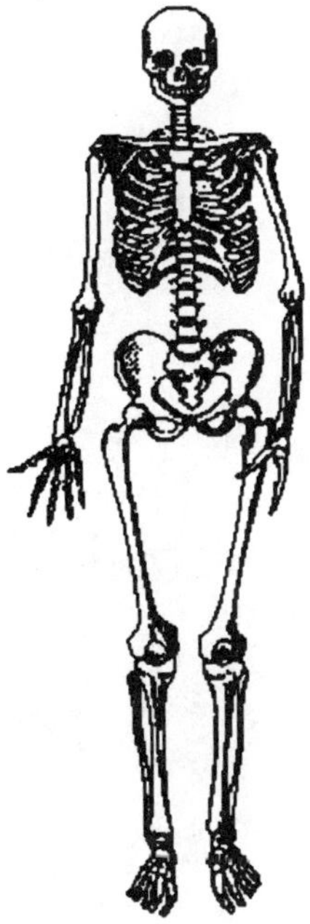

Answers to Unit 5 End-of-Unit Assessment

Word Usage

1. Answer:
   Sample answer: The Tacoma Narrows Bridge *collapsed* because the *structure* of the bridge was too weak to resist the *shear force* caused by high winds.

2. Answer:
   Sample answer: The *postmodernist* style includes *features* that serve no *practical function* but simply make the building more interesting to look at.

Short Response

3. Answer:

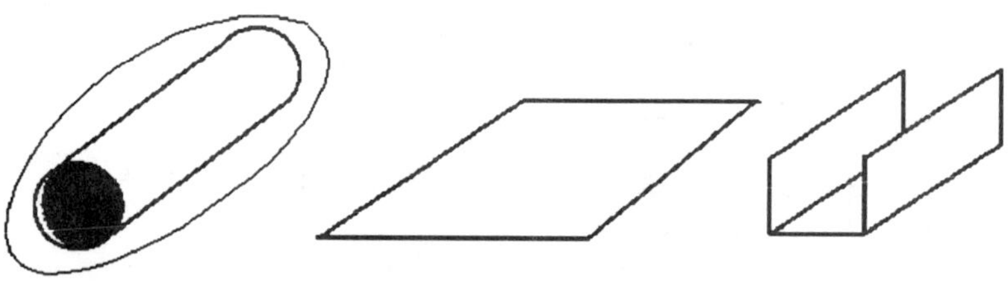

4. Answer:
   Sample answer: A truss is a framework of connected beams or bars that is built to add strength to a bridge, roof, or other structure.

Illustrative

5. Answer:
   a. Cantilevers
   b. Beams
   c. Cantilevers
   d. Beams

6. Answer:
   a. Ancient Greek architecture
   b. Modern architecture
   c. Renaissance architecture
   d. Nomadic architecture

Graphic

7. Answer:
   a. About 2 cm
   b. Compression
   c. The bridge collapsed.
   d. Sample answer: Laminate it, corrugate it, or use stronger materials.

8. Answer:

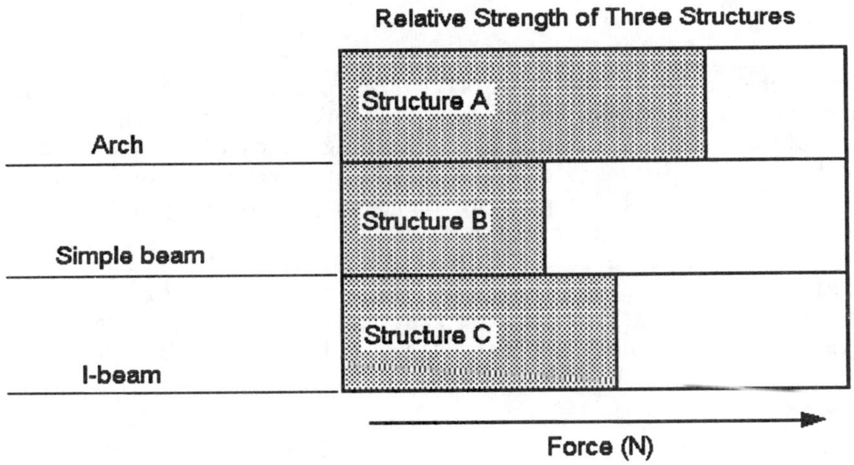

Illustrative

9. Answer:

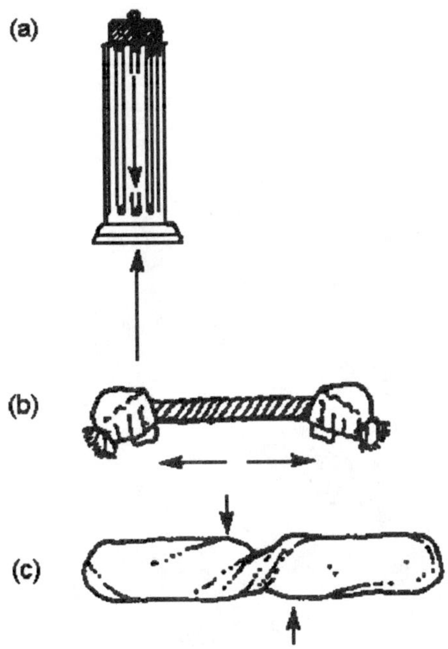

Difficulty: 3

Short Response

10. Answer:
    a. Sample answer: The bones that make up the skeletal system provide a solid framework for the shape and functions of the human body; they also protect vital organs. Ligaments and tendons solidify that structure by connecting bones at joints or by connecting muscles to bones. Muscles enable a wide range of movements, including eating, drinking, running, walking, lifting, etc. The skin provides a protective outer layer that prevents matter from passing freely into and out of the body.
    b. All of the main components of human-built structures are found in the human body. For example, leg and arm bones are beams, the spine and rib cage are trusses, the shoulder and arm connections are cantilevers, the feet and rib cage are arches, some joints are hinges, ligaments and tendons are fasteners, and the skin is a protective outer layer.

    Difficulty: 4

Unit 5 Activity Assessment

Activity Assessment

1. Bridging the Gap

Teacher's Notes

Overview
Students design a simple bridge and determine how much weight their structure can hold. They draw a sketch and build a model of their bridge.

Materials
(per activity station)
- masking tape
- a piece of poster board
- scissors
- a metric ruler
- 100 g, 250 g, and 500 g masses

Preparation
Prior to the assessment, equip student activity stations with the materials needed for each experiment.

Time Required
Each student should have 30 minutes at the activity station and 15 minutes to test the strength of the bridge.

Performance
At the end of the assessment, students should turn in the following:
- a completed Data Chart
- a poster-board bridge

Evaluation
The following is a recommended breakdown for evaluation of this Activity Assessment:
- 50% appropriate use of materials and equipment
- 50% strength of bridge

Safety Alert!

Activity Assessment

2. **Bridging the Gap**
   How does an architect decide what kind of bridge to build or how to make a bridge strong? How would you do it? Try your hand at an architectural adventure. Using the materials at your activity station, build a bridge that spans 25 cm, and then test its strength.

   Before You Begin . . .
   As you work through the tasks, keep in mind that your teacher will be observing the following:
   - how well you use the materials and equipment
   - how much weight your bridge can support

   Get ready to bridge the gap!

   Task 1 Use poster board and masking tape to design and construct your bridge. The bridge must span 25 cm, and you need to make it as strong as you can.

   Task 2 Test the sturdiness of your bridge by placing each of the different masses on it, beginning with the lightest mass. For each mass, record the results on the Data Chart. Describe what happens to the structure when the different masses rest on it.

   Task 3
   a. On the next page, draw a sketch of your bridge.
   b. Use arrows to show the direction of forces on the bridge.
   c. Use an X to show the point where the bridge failed.
   d. Indicate points on the bridge that you could improve.
   e. Explain how these improvements would make your bridge stronger.

   Safety Alert!

3.

Data Chart

Masses	Observations
100 g	
250 g	
500 g	

Sketch

Answers to Unit 5 Activity Assessment

1. Answer: Not applicable (teacher's notes)

2. Answer: Not applicable (student's notes)

3. Answer:

Data Chart

Masses	Sample Answers: Observations
100 g	The roadway surface sagged slightly (0.5 cm) and the supports (legs) pressed outward against the buttresses (books), but the bridge held.
250 g	The roadway surface sagged significantly (1.2 cm). The supporting arch began to crease and bend asymmetrically but the buttresses (books) and the bridge held.
500 g	The bridge collapsed. The crease that developed in the supporting arch under the 250 g load buckled, and the bridge collapsed. The buttresses did not move.

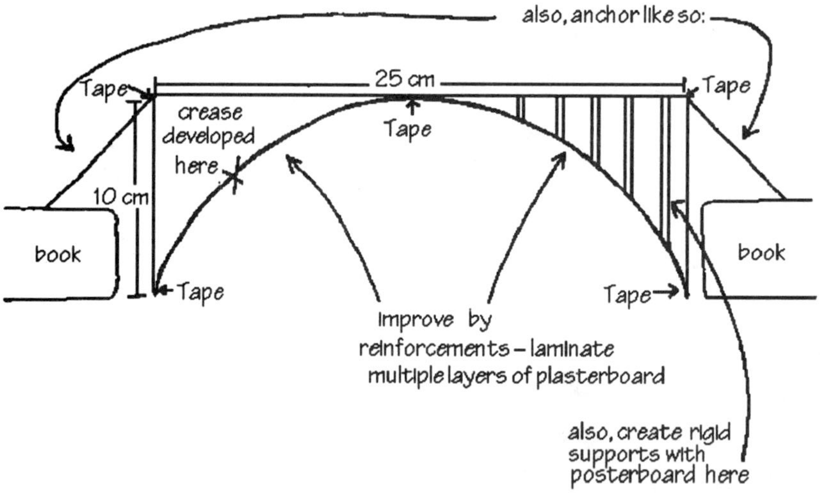

Unit 5 SourceBook Assessment

Multiple Choice

1. The Hindu temples are very plain and open, with few walls enclosing them.
   a. true b. false

2. One of the arts of the Moslems was to make miniature gardens that were representations of their surrounding world.
   a. true b. false

3. In the moist, wet areas around the Mississippi River, the early inhabitants built their homes out of adobe brick.
   a. true b. false

4. African architecture included the development of urban areas with marketplaces, streets, and other city structures.
   a. true b. false

5. The magnificent Taj Mahal was built to be a
   a. city hall. b. sports arena. c. public bath house. d. tomb.

6. Symmetrical design is characteristic of __________ architecture.
   a. Hindu b. Chinese c. Islamic d. Aztec

7. The gently upcurved roof line of many shrines and temples is a typical architectural style of the
   a. Japanese. b. Incas. c. Hindus. d. Chinese.

8. The Great Pyramid in Egypt has a (larger, smaller) base than the Pyramid of the Sun found near Mexico City. (Circle the correct word.)

9. The Maya were known for their
   a. architecture.
   b. writing system.
   c. calendar.
   d. mathematics.
   e. All of the above
   f. a and c only

10. Which of the following was not one of the distinctive architectural cultures in Central and South America?
    a. Incan b. Aztec c. Samoan d. Mayan

11. The Incas made their homes and palaces out of stones. How did they hold the stones together?
    a. with mortar
    b. They cut the stones accurately so that they fit together.
    c. with cement
    d. with nails

12. Building high walls around cities was a common feature of ___________ architecture.
    a. African b. Mexican c. Japanese d. Incan

13. Some of the most important structures built in the early years of Islamic culture were
    a. religious buildings.
    b. civic centers.
    c. city courts.
    d. sports arenas.

Short Response

14. Describe one difference between a tepee and a wigwam.

____

____

____

____

15. Why were tepees the best kind of home for the Sioux Indians?

____

____

____

____

16. Some of the early North American cultures built huge ceremonial mounds around the Mississippi and Ohio Rivers. Name one purpose of these huge mounds.

____

____

17. Describe how architectural styles originated due to the influence of climate or culture. Give examples.

____

____

____

____

Answers to Unit 5 SourceBook Assessment

Multiple Choice

1. Answer: b. false

2. Answer: b. false

3. Answer: b. false

4. Answer: a. true

5. Answer: d. tomb.

6. Answer: b. Chinese

7. Answer: a. Japanese.

8. Answer: smaller

9. Answer: e. All of the above

10. Answer: c. Samoan

11. Answer: b. They cut the stones accurately so that they fit together.

12. Answer: a. African

13. Answer: a. religious buildings.

Short Response

14. Answer:
    Sample answer: A tepee is a portable, cone-shaped tent covered with animal skins. A wigwam is a large, rounded tent or structure covered with leaves, tree bark, or reed mats. It is not as easy to transport as a tepee.

15. Answer:
    The Sioux Indians were nomadic; they traveled to follow the buffalo herds. Tepees were easy to disassemble and move to their next location.

16. Answer: They served as burial mounds and as bases for temple structures.

17. Answer:
Answers may vary but should indicate that the student understands the relationship between culture or climate and architecture.

Unit 5 Extra Assessment Items

Word Usage

1. Use the following words in one or two sentences to show how they are related: structure, fail, design, and materials.

2. For each of the following terms, write a sentence that explains the effect that each of these forces has on an object:
   a. tension

   b. compression

   c. shearing

3. Show your understanding of the following terms as they relate to structures and design by using them in one or more sentences:

   a. preservation

   b. urban restoration

Correction/Completion

4. Draw lines to connect the periods of architecture on the left with the correct characteristics on the right. Then number the periods of architecture so that they are in the correct order.

Periods of architecture	Characteristics
_____ Early civilization	• classic structures, including aqueducts and the Colosseum
_____ Modern civilization	• ornate domes, churches, and other buildings, including the dome of the Cathedral of Florence and St. Peter's Basilica
_____ Ancient Rome	• pillars and columns like those found in the Acropolis and Parthenon
_____ Renaissance	• buildings made of reinforced concrete and steel, including skyscrapers and stadiums
_____ Medieval Europe	• permanent stone buildings such as the Great Pyramid of Cheops
_____ Ancient Greece	• huge cathedrals such as Notre Dame in Paris

Short Essay

5. Why do you think that modern buildings do not have the intricate designs of the buildings made during the Renaissance (1300s–1600s)?

____

____

____

____

____

____

____

____

____

____

____

____

6. Considering the shape and materials used, describe why an igloo and not a tepee would be used in the Arctic.

Short Response

7. Forces of nature and natural disasters are unpredictable. Describe the effects that the following occurrences might have on a structure:
   a. hurricanes

   b. earthquakes

   c. floods

Illustrative

8. Consider the three shapes shown below.

   Select one of these shapes and describe why you think it would support the greatest mass.

   ____

   ____

   ____

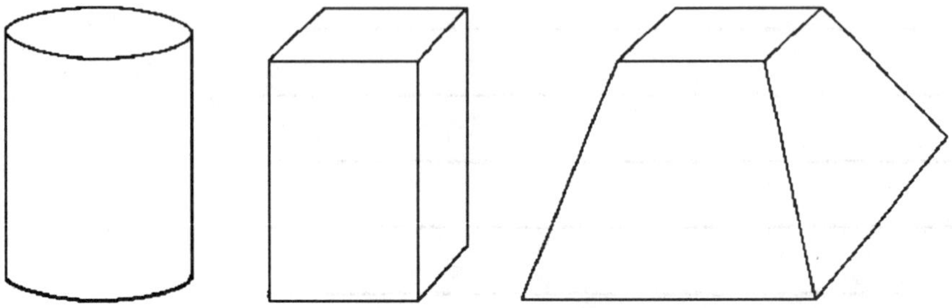

9. A rope holding up a load could stretch and twist at the same time. Describe the action of the force on a rope if the load is
   a. decreased.

   ____

   ____

   ____

   b. increased.

   ____

   ____

   ____

10. Lee used a meter stick to illustrate her understanding of beams and cantilevers. Look at the demonstrations on the next page. Was she right? Explain your reasoning.

   ____

   ____

   ____

11. Niall and Sheila are having an argument. Sheila has supported a book on a table using a brick, as shown below. She claims that this is an example of a cantilever. Niall says that it is an example of a beam. Who is right? Explain your reasoning.

____

____

____

Short Response

12. In one or two sentences, explain
    a. why an I-beam is stronger than a simple beam.

____

____

____

    b. why an arch is stronger than a simple beam.

____

____

____

13. What role does a buttress play in an arched bridge?

____

____

____

14. In a stone archway, the center stone at the top is called the *keystone* because of its importance. Why do you think it is so important?

____

____

____

15. If you were designing a bridge to be built across the Grand Canyon, would you select an arched bridge, a cantilever bridge, or a suspension bridge? Explain.

____

____

____

____

Graphic

16. Below are the results of Dr. Jones's experiment on the stretching of a nylon fiber.

Force used to stretch the fiber (N)	Stretch of the 20 cm long, 2.4 mm diameter fiber (mm)
0	0.0
10	1.4
20	3.0
30	4.6
40	6.1

a. Graph this data. Let the horizontal axis represent force and the vertical axis represent stretch. Represent each data pair with a dot on the graph. Draw a straight line that is close to most of the points.
b. Use the graph to answer the following questions: How much would the fiber have stretched if Dr. Jones had applied a force of
 i. 25 N? _________
 ii. 5 N? _________
 iii. 50 N? _________
c. If the piece of nylon fiber that Dr. Jones stretched had been shorter, would the amounts of stretching have been greater, less, or the same?
d. If the thickness of the fiber used had been greater, would the amounts of stretching have been greater, less, or the same?

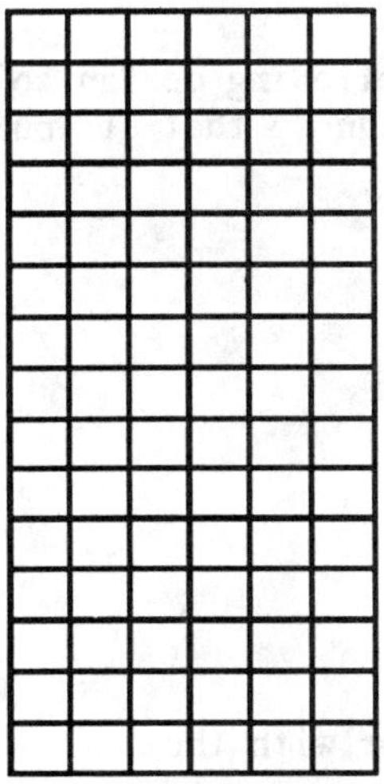

Illustrative

17. Tara added masses to the bridges shown in the diagrams below. She continued to add masses until each bridge collapsed. Label the kind of structure used in each diagram, and circle the one that supported the greatest weight.

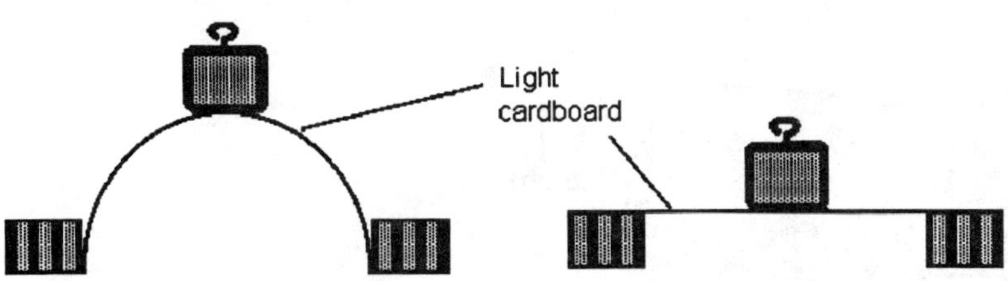

18. Identify the kind of bridge shown in each illustration.

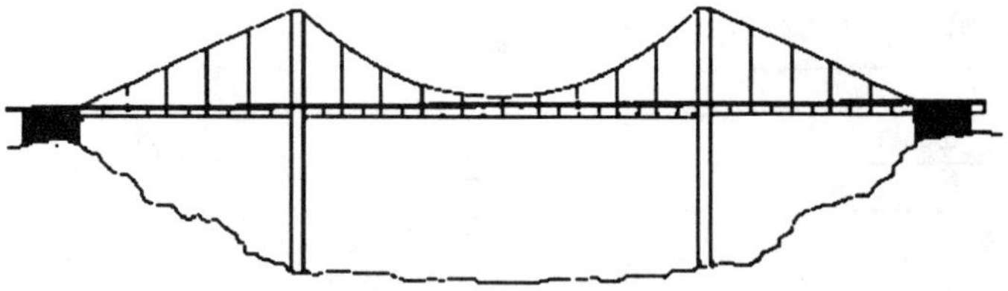

a. ____________________

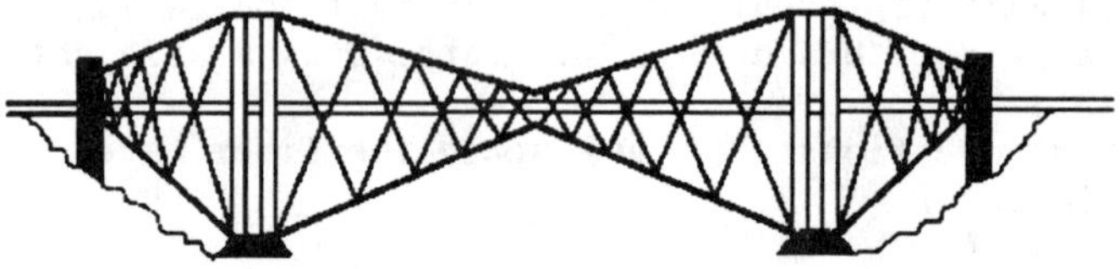

b. ____________________

c. ____________________ d. ____________________

19. In a contest, students were asked to increase the strength of the following design so that it would support twice as much weight as shown. Redraw the design so that it would be stronger.

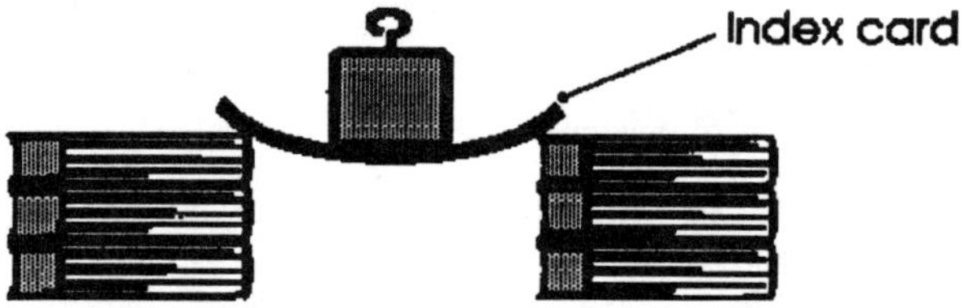

20. You are given soft modeling clay that has been rolled into a cylinder with the dimensions shown.
   a. Sketch what would happen if you placed tensile forces on the clay. Draw arrows to indicate the direction of the forces.
   b. Sketch what would happen if you placed shear forces on the clay. Draw arrows to indicate the direction of the forces.
   c. Sketch what would happen if you placed compressive forces on the clay. Draw arrows to indicate the direction of the forces.

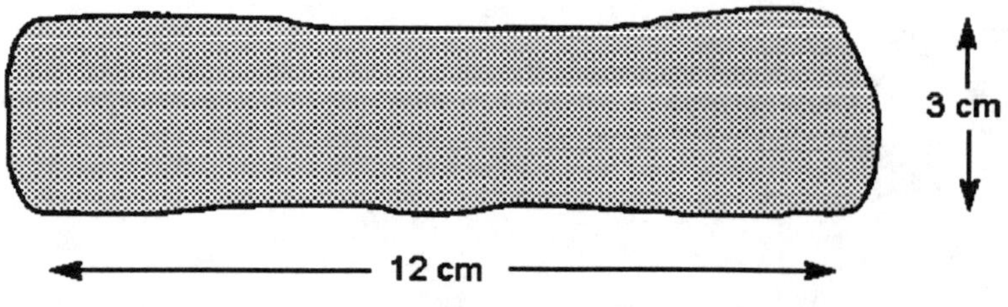

21. Consider the cardboard bridge shown.
Choose one of the following techniques to strengthen the bridge, and sketch the result of applying that technique. You may use any additional materials you need in your new design.
a. building a truss
b. using corrugation
c. rebuilding the bridge as an arch

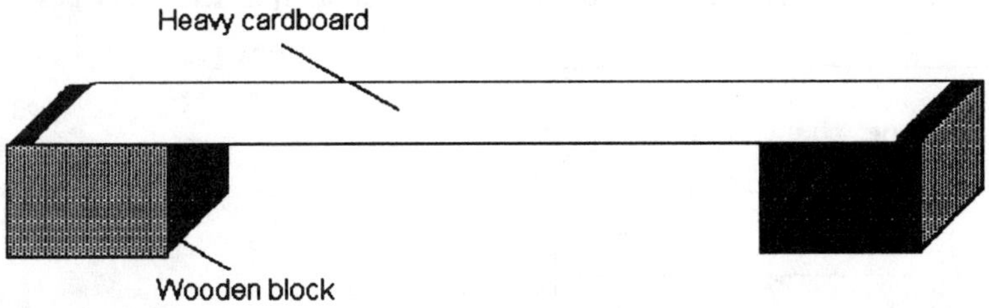

Performance Task

22. **Task:** Suppose you want to find out which wire makes a stronger fishing line: wire made of nylon or wire made of copper. What procedure would you follow to determine which is the stronger material?

Procedure:

____

____

____

____

____

____

____

23. **Materials**: a clamp, a metal strip, a metric ruler, standard masses or a force meter, two different colored pencils

Task 1: Clamp the metal strip to the edge of a table. Measure the amount of metal that extends beyond the table. Next, apply a standard mass or force to the end of the strip to determine how much the metal strip bends. Record your results in the table below. Then increase the force, and again, record your results. Use at least four different forces. Record your results in the table below, and then graph the results.

Force applied to metal strip that extends ______ from the table	Distance the metal strip bends

Task 2: Repeat Task 1, but this time extend the metal strip a shorter distance from the end of the table. Record your data in the chart below. Then use a different colored pencil to plot your results on the graph you used for Task 1.

Force applied to metal strip that extends ______ from the table	Distance the metal strip bends

Task 3: Analyze your data. What can you conclude?

____

____

____

____

____

24. On a separate piece of paper, sketch a bridge that exists in your community. Identify the type of bridge, what forces it experiences, and how it is designed to resist these forces.

25. On a separate piece of paper, sketch a home or building in your community that has features that remind you of an earlier period in history. Explain.

26. On a separate piece of paper, design plans to improve some aspect of your community. Include a sketch of the way the structure or setting exists in the community now as well as a sketch detailing your plans for how the structure or setting could be improved.

Answers to Unit 5 Extra Assessment Items

Word Usage

1. Answer:
Sample answer: When a *structure* such as a bridge *fails*, engineers try to determine whether the problem was caused by faulty *design*, inferior *materials*, or some other problem.

2. Answer:
a. Sample answer: *Tension* is a force that causes a material to stretch.
b. Sample answer: *Compression* is a force that causes a material to press together into a smaller space.
c. Sample answer: *Shearing* is a force that causes a material to tear or be cut apart.

3. Answer:
a. Sample answer: The *preservation* of historic buildings allows a city to keep some of its original architecture.
b. Sample answer: Through *urban restoration*, we can clean up run-down areas, restore old buildings, and design new structures to resist deterioration over long periods of time.

Correction/Completion

4. Answer:

Periods of architecture	Characteristics
1 Early civilization	classic structures, including aqueducts and the Colosseum
6 Modern civilization	ornate domes, churches, and other buildings, including the dome of the Cathedral of Florence and St. Peter's Basilica
3 Ancient Rome	pillars and columns like those found in the Acropolis and Parthenon
5 Renaissance	buildings made of reinforced concrete and steel, including skyscrapers and stadiums
4 Medieval Europe	permanent stone buildings such as the Great Pyramid of Cheops
2 Ancient Greece	huge cathedrals such as Notre Dame in Paris

Short Essay

5. Answer:
Answers will vary but will probably refer to the differences in the cultural preferences of the people living during the Renaissance and of those living now.

6. Answer:
Cold temperatures, high winds, and a lack of trees would make the tepee unsuitable for the Arctic. The tepee would not keep out snow as well as an igloo, it could blow over in a heavy wind, and it is made from materials that are rare in the region. An igloo, on the other hand, is made from materials that are common to the area. Those materials are much more likely to withstand the harsh conditions of the Arctic. The shape of the igloo, a dome, is also better suited to the harsh conditions of the Arctic because it is much stronger than the shape of a tepee.

Short Response

7. Answer:
   Sample answers:
   a. Hurricanes have extremely high winds and can cause very high tides and flooding. These storms can blow out windows, rip off roofs, and destroy the walls of houses and other structures.
   b. Earthquakes can cause the surface of the Earth to shake and cracks to form. This could damage the foundation of a building, causing it to fall down.
   c. Floods can cause a great deal of water damage. They also transport great quantities of solid material, which may damage the supporting structures of a building.

Illustrative

8. Answer:
   The object on the right has a wider base which distributes the load over a larger area. It is also more massive.

9. Answer:
   a. When the load is decreased, the tension and twisting action of the forces decrease. The length of the rope will shorten.
   b. When the load is increased, the tension and twisting action of the forces increase. The rope will stretch further and twist.

10. Answer:
    Lee is correct. The cantilever is fixed or supported at one end, while the beam is laid across two supports.

11. Answer: Sheila is correct—the book is only supported on one end, so it is a cantilever.

Short Response

12. Answer:
    a. The force applied to an I-beam is distributed more evenly and over a larger area than the force exerted on a simple beam of the same size.
    b. The force applied to an arch is transferred to the thick walls that support the arch, while the same force applied to a simple beam affects only the beam itself.

13. Answer:
    A buttress holds the arch in place and receives the force of the load placed on the arch.

14. Answer:
    Without a keystone, an arch would collapse. The keystone supports the horizontal forces at the top of the arch.

15. Answer:
    Sample answer: I would use a suspension bridge because suspension bridges are able to span large distances with the least amount of material.

Graphic

16. Answer:
    a. See graph.
    b. i. About 3.8
       ii. About 0.7
       iii. About 7.5
    c. The stretch would be shorter since the amount of total stretch is proportional to both the pulling force and the original length. (For instance, you cannot stretch a 1 m rope 10 m without it breaking. However, 1000 m of rope will easily stretch 10 m.
    d. The stretch would be less because the fiber would have more resistance to the pulling force.

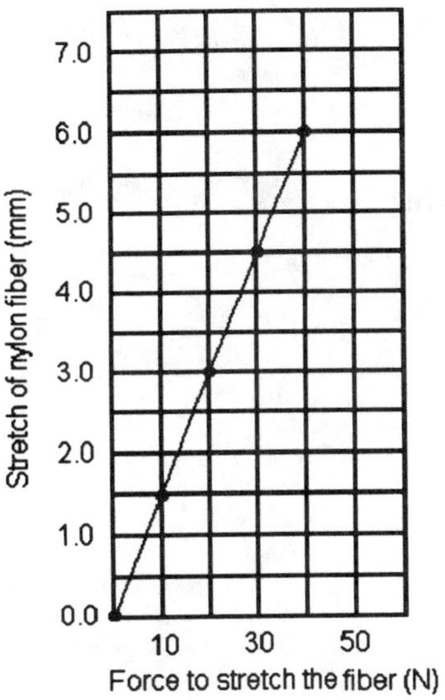

Illustrative

17. Answer:

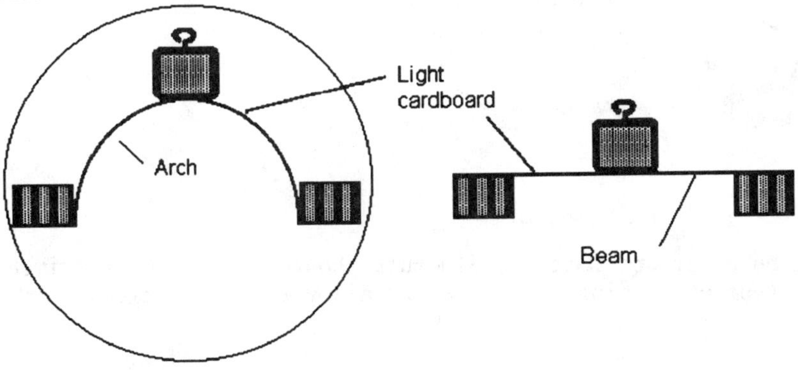

18. Answer:
    a. Suspension bridge
    b. Cantilever bridge
    c. Arch bridge
    d. Beam or truss bridge

19. Answer:
    In the students' sketches, the card could be laminated, the design could be changed into an arch, or support beams could be added.

20. Answer:

a.

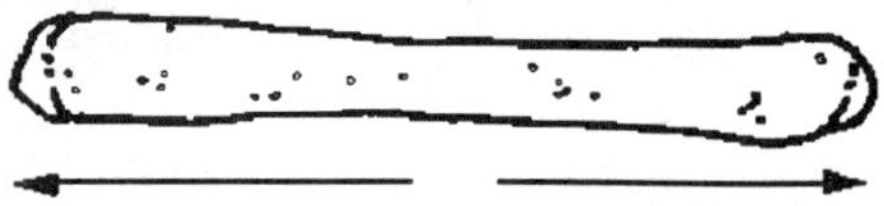

b.

c.

21. Answer: Answers will vary but could include one of the following designs.

a.

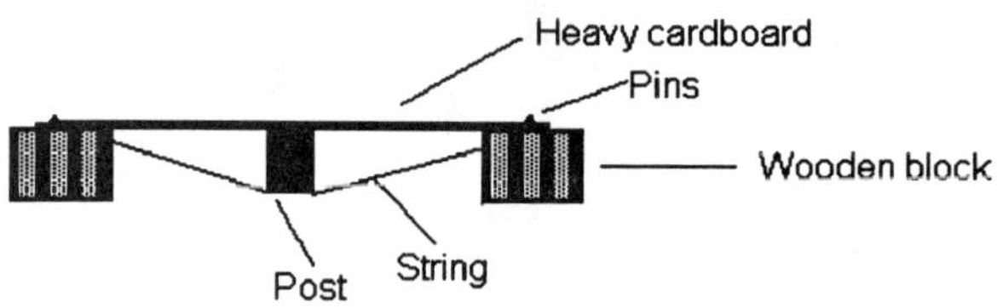

b.

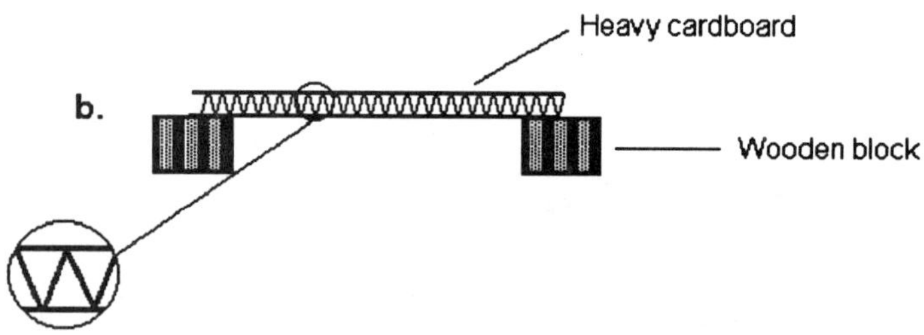

c.

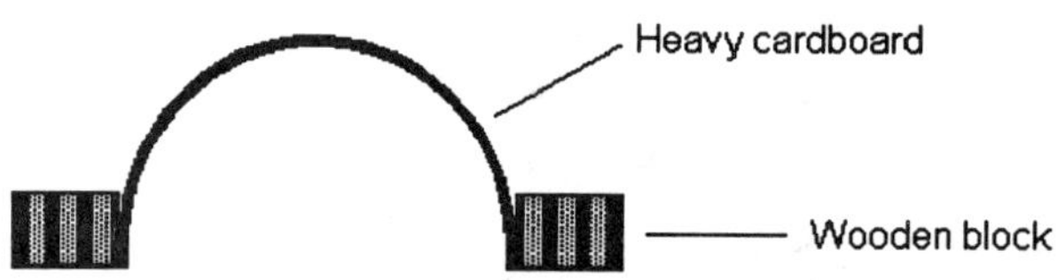

Performance Task

22. Answer:
Answers will vary but should be clear and concise. Students should design an experiment to test the strength of each type of fishing line, keeping all variables constant except the type of fishing line.

23. Answer:
Answers will vary, but should be supported by data expressed on the graph and in the tables.

24. Answer:
Answers will vary, but students should demonstrate an understanding of their chosen bridge type and the way it responds to forces.

25. Answer:
Answers will vary, but students should demonstrate an understanding of an historical architectural style or influence.

26. Answer:
Answers will vary, but students should demonstrate an understanding of how land-use planning can benefit a community.

Chapter 16 Assessment

Correction/Completion

1. Complete the following statements:
   a. Mountains are formed when . . .

   ____

   ____

   b. Most earthquakes and volcanoes occur where . . .

   ____

   ____

Short Response

2. Arrange the following items in sequence so that they tell the story of a mountain:
   a. Wind, rain, and other events can wear down the uplifted rock and soil to form a hill.
   b. The crustal plates are in continuous motion.
   c. As the crustal plates collide and fold, the Earth's surface can be uplifted.
   d. The Earth's crust is made up of large plates.

   Sequence: ____________________

Illustrative

3. Mark *E*s and *V*s on the following map to show where most earthquakes and volcanoes probably occur:

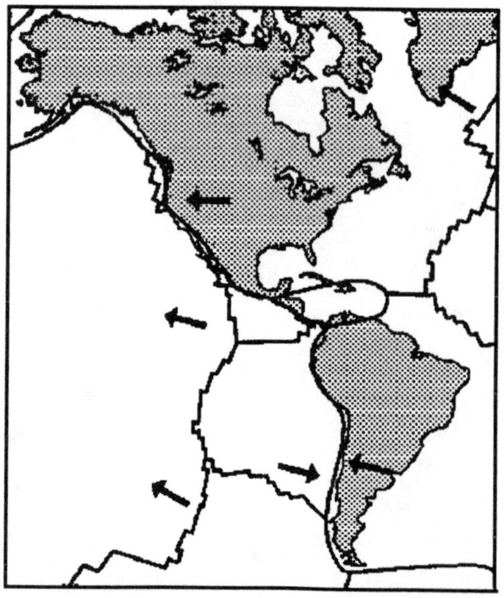

Numerical Problem

4. During an earthquake, about 100 J of energy is released from each cubic meter of rock. Suppose a fault is 200 km long, extends 50 km downward, and bends streets as far away as 20 km on either side of the fault.
   a. How many cubic meters of rock would be under strain?

   ____

   b. How much energy would be released when the rocks break the friction lock?

   ____

Illustrative

5. Explain what probably caused this rock formation.

____

____

____

____

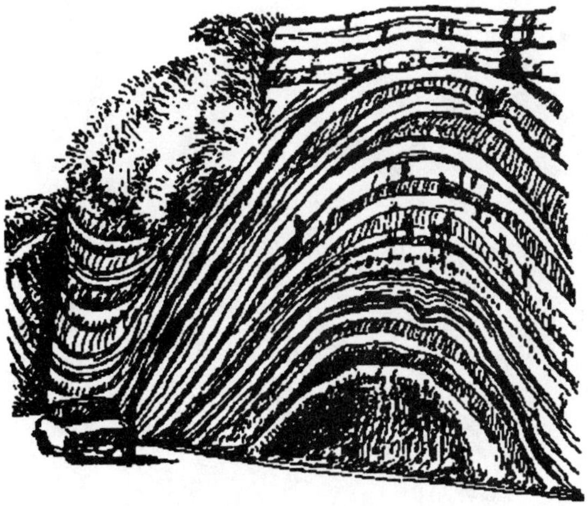

Short Essay

6. An island tripled in size in 1 hour! Explain how this probably happened.

Answers to Chapter 16 Assessment

Correction/Completion

1. Answer:
   a. . . . tectonic plates collide and the crust of the Earth buckles and is forced upward.
   b. . . . plates are pushing against each other or moving away from each other.

Short Response

2. Answer: Sequence: d, b, c, a

Illustrative

3. Answer:

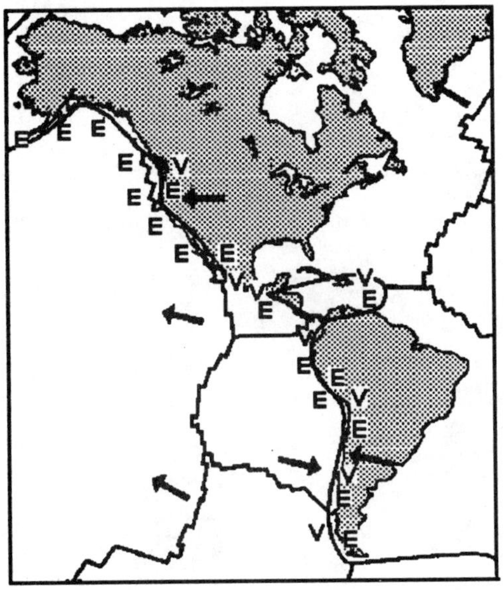

Numerical Problem

4. Answer:
   a. $200 \text{ km} \times 50 \text{ km} \times 40 \text{ km} = 400{,}000 \text{ km}^3 \times \dfrac{10^9 \text{m}^3}{1 \text{ km}^3} = 4 \times 10^{14} \text{ m}^3$

   b. $4 \times 10^{14} \text{ m}^3 \times \dfrac{100 \text{ J}}{1 \text{ m}^3} = 4 \times 10^{16}$ J of energy

Illustrative

5. Answer:
Sample answer: The rock formation is probably the result of folding that occurred as two tectonic plates pushed against each other. When this happened, the rock layers, which had been flat, curved into the pattern shown.

Difficulty: 3

Short Essay

6. Answer:
Sample answer: The island was probably originally created by a volcano. It most likely increased in size because the volcano erupted again. This occurred when two plates separated and a crack in the ocean floor opened, causing lava to flow to the surface. The lava cooled when it came in contact with the air and surrounding water, and the island increased in size.

Difficulty: 4

Chapter 17 Assessment

Word Usage

lava	molten	intrusive	erosion	deposition	crystallization
heat	cementation	compression	cooling	extrusive	volcanic
weathering	basalt	plutonic	granite	lithification	pressure

1. Pick at least two of the words from the list that are related to the word *igneous*, and then use those words in a sentence to show that you know how they are related.

____

____

____

2. Pick at least two of the words from the list that are related to the word *sedimentary*, and then use those words in a sentence to show that you know how they are related.

____

____

____

3. Pick at least two of the words from the list that are related to the word *metamorphic*, and then use those words in a sentence to show that you know how they are related.

____

____

____

Illustrative

4. Jonas has written the following story to help a third-grade student better understand the rock cycle. Help Jonas by creating a simple diagram to illustrate his story. Also provide a specific example of rock for each stage in the cycle. Record your answer on the back of this paper.

   During a volcanic eruption, some rocks were formed. Over many years, bits and pieces of those and other rocks were washed into a river. Eventually, some of them settled to the bottom and became compacted. They remained there under tremendous pressure until a tectonic plate moved under the deposits, melted them, and forced the molten rock to the surface.

Short Response

5. Jerome believes that the Earth was always basically the same as it appears to be today. You want to convince him that he is wrong. Make a list of the evidence that you will present to Jerome to prove that the Earth has changed over time.

Short Essay

6. Obsidian is an interesting igneous rock. It is smooth, black, and shiny. Much like glass, it has no grain. Native Americans used it to make knives and arrowheads that were sharper than steel. Discuss how obsidian forms, and explain how that process gives it properties that make the rock useful for creating knives and other cutting tools.

Answers to Chapter 17 Assessment

Word Usage

1. Answer:
   Sample answer: *Basalt*, the most common type of *extrusive* igneous rock, is formed from *molten lava*.

2. Answer:
   Sample answer: Sedimentary rocks are formed as a result of *deposition* and *lithification*; deposition occurs because of *weathering*, and lithification occurs because of *cementation* or *crystallization*.

3. Answer:
   Sample answer: Metamorphic rocks are formed when *heat* and *pressure* change the appearance and characteristics of existing rocks.

Illustrative

4. Answer: Sample answer:

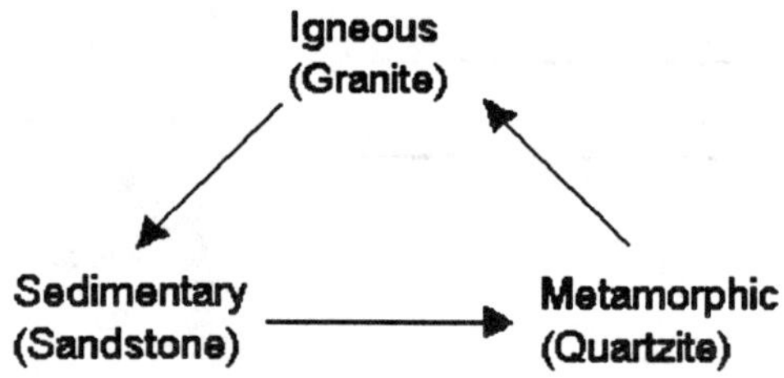

Short Response

5. Answer:
   Possible answers include fossil evidence, evidence of continental drift and plate tectonics, evidence of mountain-building, evidence of vast sedimentary deposits, folding and faulting, and large displacements of land associated with earthquakes and volcanic activity. Students may also mention various types of erosion, such as water erosion, wind erosion, tidal erosion, and erosion through glaciation, that have shaped the surface of the Earth.

   Difficulty: 3

Short Essay

6. Answer:
Sample answer: Obsidian is an igneous rock formed in volcanoes. It must cool extremely quickly because otherwise it would have grains, and the slower it cooled, the larger the grains would be. Native Americans probably broke chunks of obsidian along fractures, resulting in the extremely sharp, curved edges that made useful cutting tools.

   Difficulty: 4

Chapter 18 Assessment

Short Response

1. For each organism below, indicate whether it has a good, fair, or poor chance of becoming a fossil, and explain why.

   a. a tree buried in a mud slide

   ____

   ____

   ____

   b. a leaf floating in a stream

   ____

   ____

   ____

   c. a beached whale

   ____

   ____

   ____

2. Supply the missing heading for each list.

________	________	________
Cenozoic	Cretaceous	Pliocene
Mesozoic	Jurassic	Miocene
Paleozoic	Triassic	Paleocene

3. Number the following events from 1 to 6 as they would appear on a relative time scale.

   ___ World War II occurred.

   ___ Dinosaurs roamed the Earth.

   ___ The first organism crawled out of water onto land.

   ___ Rock-and-roll music became popular.

   ___ Humans discovered fire.

   ___ President Clinton took office.

Illustrative

4. The following are the notes that a paleontologist, Dr. Henry Bones, jotted down when looking at an archeological site. Sketch a representation of what he might have seen.
   - Two animals were involved.
   - The animals were moving in the same direction.
   - Each animal walked on its hind two legs.
   - Each footprint shows four toes.
   - Both sets of footprints are similar in appearance, but one set is much larger than the other.
   - The footprints of the small animal are very close to that of the larger animal, but they never intersect.

   What story might these footprints tell?

   ____

   ____

   ____

   ____

   ____

   ____

Short Response

5. An amateur paleontologist claims to have made a startling discovery—dinosaur tracks and human tracks fossilized in the same rock layer. This implies that dinosaurs and humans lived on the Earth at the same time. Which of the following questions would a scientist ask if he or she were testing the paleontologist's claims?
   a. What kind of dinosaur is it? Was the human male or female?
   b. How big are the dinosaur footprints compared to the human footprints?
   c. How old is the rock layer? How old are each of the fossilized footprints within the rock layer?
   d. What was the Earth like in this location during the Precambrian era?

Numerical Problem

6. Potassium-40 is a radioactive isotope that scientists use to date rocks and minerals. It has a half-life of approximately 1,300,000,000 years. If you had a kilogram of potassium-40, how much of it would still be radioactive in 5,200,000,000 years? Show your work in the space provided.

Answers to Chapter 18 Assessment

Short Response

1. Answer:
   a. A fair chance. Trees usually rot, but the mud might protect it.
   b. A poor chance. Leaves decompose quickly, especially in water.
   c. A good chance. The bones are hard and may become buried in the sand.

2. Answer:

Era	Period	Epoch

3. Answer:
   4 World War II occurred.
   2 Dinosaurs roamed the Earth.
   1 The first organism crawled out of water onto land.
   5 Rock-and-roll music became popular.
   3 Humans discovered fire.
   6 President Clinton took office.

Illustrative

4. Answer:
   Illustrations will vary but should reflect a careful reading of the notes. Sample illustration below.

   Sample answer: An adult dinosaur led a baby dinosaur from one place to another.

Short Response

5. Answer:
   c. How old is the rock layer? How old are each of the fossilized footprints within the rock layer?

Numerical Problem

6. Answer:
   The quantity of potassium-40 is cut in half every 1,300,000,000 years.
   1 kg becomes 0.5 kg at 1,300,000,000 years
   0.5 kg becomes 0.25 kg at 2,600,000,000 years
   0.25 kg becomes 0.125 kg at 3,900,000,000 years
   0.125 kg becomes 0.0625 kg at 5,200,000,000 years
   So there is 0.0625 kg of potassium-40 left over from 1 kg after 5,200,000,000 years.

Unit 6 End-of-Unit Assessment

Word Usage

1. Beside each item in list A, write at least three related words from list B. Then write a sentence for each item in list A, including at least one of the related words you chose from list B.

List A

a. volcano: ______________________________

____

____

____

b. igneous: ______________________________

____

____

____

c. plate tectonics: ______________________________

____

____

____

d. geologic time scale: ______________________________

____

____

____

List B

basalt	Precambrian	extrusive	era	transform
crystal	intrusion	dike	divergent	Mesozoic
pressure	heat	fossil	magma	cooling
granite	continental	intrusive	collide	lava
sill	molten	sliding	Paleozoic	convergent

Short Response

2. To demonstrate how the elastic rebound theory explains the origin of earthquakes, Jan's teacher performed the following demonstration using an elastic band and plastic rulers.

   You be the teacher and explain the connection between the demonstration and the elastic rebound theory.

   ____

   ____

   ____

   ____

   ____

   ____

   ____

   ____

   a. The elastic band is held loosely.

   b. The elastic band is stretched.

   c. The elastic band is cut.

3. Draw lines to match the event on the left with the appropriate era on the right.

Event	**Era**
Dinosaurs flourished.	Cenozoic
The first fish appeared.	Mesozoic
Humans became abundant.	Paleozoic

4. Aaron was shown a fossil of a leaf imprint. "That's probably a sedimentary rock," he said. Is he right? Why or why not?

5. This graphic shows part of the geologic time scale.

a. Give an example of an era.

b. Give an example of a period.

c. Dinosaurs died out during the Cretaceous part of the geologic time scale. Were they alive during the Tertiary or the Jurassic part?

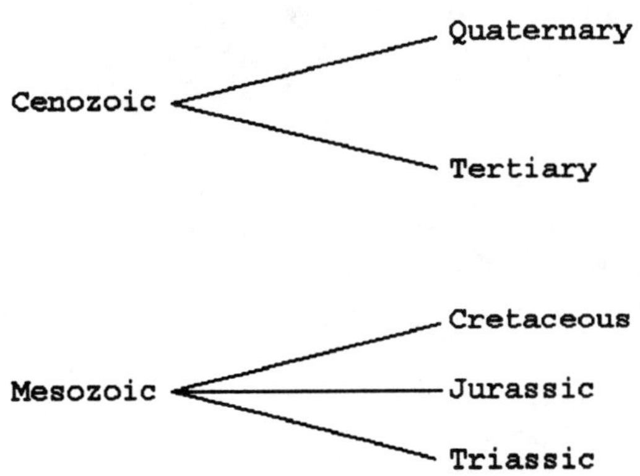

6. The following statements describe how fossils are formed, but the sequence shown is out of order. Number the statements to show their correct order.

___ a. Silt forms sedimentary rock over time.

___ b. A living thing dies.

___ c. Erosion causes a fossil-bearing rock to be exposed.

___ d. A once-living thing becomes covered with silt and sand.

Illustrative

7. One possible way of representing the rock cycle is shown in the diagram below. Give a possible explanation for the changes represented by each of the arrows.

Arrow 1: ______________________________

____

____

Arrow 2: ______________________________

____

____

Arrow 3: ______________________________

____

____

Arrow 4: ______________________________

____

____

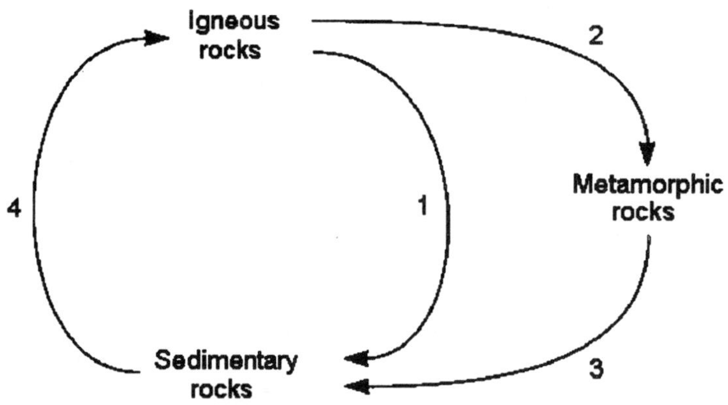

8. Complete the illustration below. When all of the labels are correctly placed, the picture will show the rock cycle.

Match the following letters to the appropriate part of the illustration:

a. an oceanic plate
b. a continental plate
c. magma
d. igneous rock
e. metamorphic rock
f. sedimentary rock
g. high-temperature, low-pressure zone
h. high-temperature, high-pressure zone
i. low-temperature, high-pressure zone

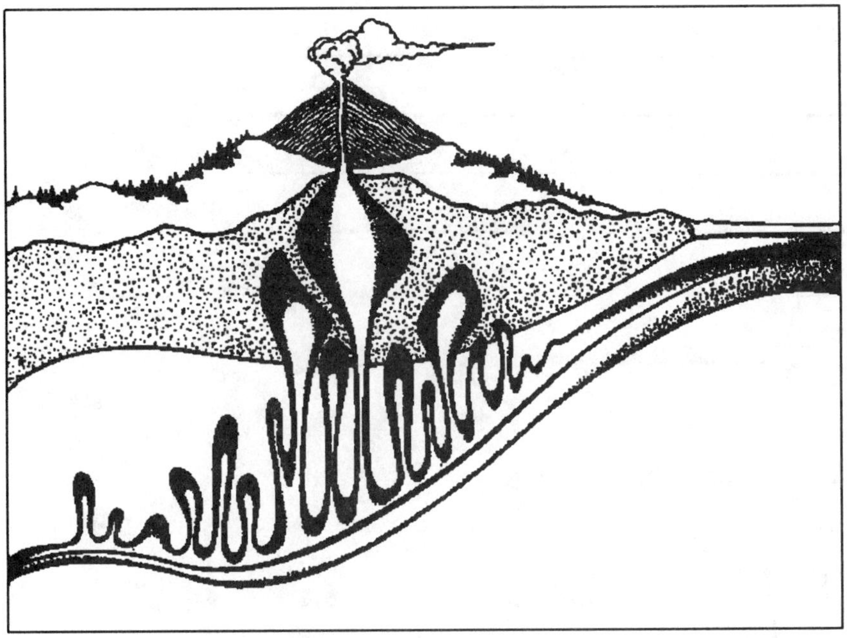

Short Essay

9. Examine the diagram of the cliff below. Use examples from the diagram to illustrate your understanding of relative time.

___

___

___

___

___

___

___

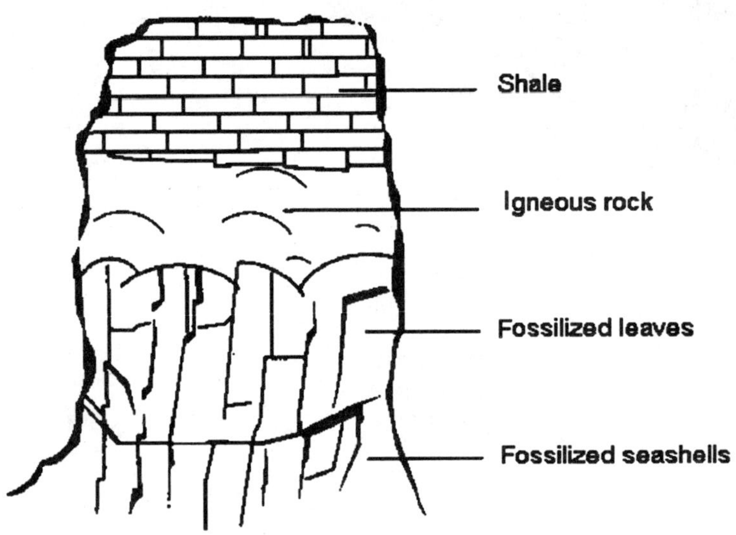

Graphic

10. The data below show the average grain size and depth of formation for a series of igneous rock samples. Use this information to complete the tasks that follow.

Average grain size (mm)	Depth of formation (km)
0.5	0.0
1.0	0.5
1.5	0.8
2.0	0.9
2.5	1.0
3.0	1.1
3.5	1.3
4.0	1.6
4.5	1.9
5.0	2.5
5.5	3.0
6.0	3.2

a. Construct a graph using the data in the chart above.

b. Use your graph to explain how the average grain size is related, or is not related, to the depth at which the igneous rock formed.

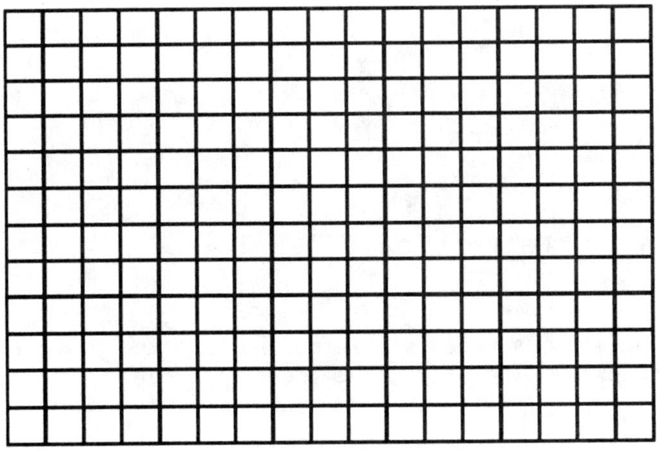

Answers to Unit 6 End-of-Unit Assessment

Word Usage

1. Answer:
   a. Sample answer: volcano: heat, pressure, basalt, magma, collide, lava
      When plates *collide*, the energy of the movement *heats* the rock until it melts into *magma* and erupts as a volcano.
   b. Sample answer: igneous: heat, basalt, molten, cooling, granite, crystal, intrusive, extrusive, sill, dike, intrusion
      Igneous rocks that *cool* slowly form *intrusive* rocks with large *crystals*.
   c. Sample answer: plate tectonics: collide, continental, convergent, divergent, sliding
      In plate tectonics, *convergent* boundaries are areas where plates are pushing together, while *divergent* boundaries are areas where plates are *sliding* away from each other.
   d. Sample answer: geologic time scale: Precambrian, Mesozoic, Paleozoic, fossil, era
      The geologic time scale is divided into *eras*; during the *Precambrian* era, only the most primitive life-forms existed on Earth.

Short Response

2. Answer:
   Sample answer: The rulers represent the rocks along the fault line. When stretched, the elastic band represents the stored energy as the rocks are deformed. When the elastic band is cut, the energy is released, and the rulers spring back to their original shape. Deformed rocks rebound in a similar way after energy is released during an earthquake.

3. Answer:

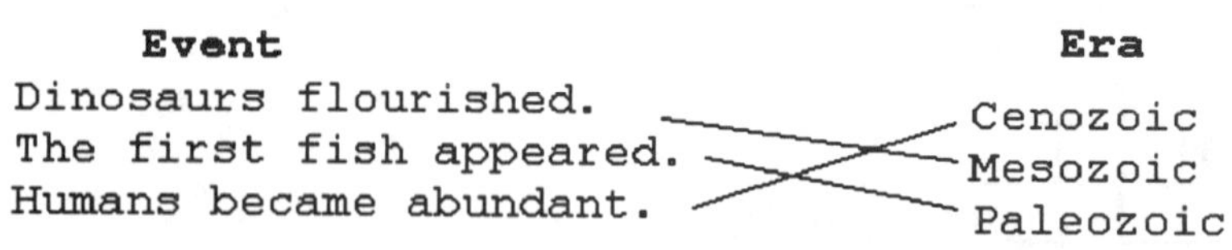

4. Answer:
   Aaron is probably right. Metamorphic and igneous rocks are formed by heat and pressure that usually destroy any once-living material. Therefore, the rock is probably sedimentary.

5. Answer:
   a. Cenozoic or Mesozoic
   b. Quaternary, Tertiary, Cretaceous, Jurassic, or Triassic
   c. Jurassic

6. Answer:
   3 a. Silt forms sedimentary rock over time.
   1 b. A living thing dies.
   4 c. Erosion causes a fossil-bearing rock to be exposed.
   2 d. A once-living thing becomes covered with silt and sand.

Illustrative

7. Answer:
   Sample answers:
   Arrow 1: Weathering breaks down the igneous rocks; these rock particles are carried to the ocean, where lithification occurs and turns the rock particles into sedimentary rock.

   Arrow 2: Igneous rocks near intrusions of molten rock are chemically altered and become metamorphic rocks, or igneous rocks are buried deep within the Earth and subjected to extreme heat and pressure, causing them to change into metamorphic rocks.

   Arrow 3: Metamorphic rock is weathered and carried to the ocean; the pressure of the water causes lithification to occur and turns the rock pieces into sedimentary rock.

   Arrow 4: Sedimentary rocks can be buried until the Earth's heat melts them. This melted rock cools into igneous rock.

   Difficulty: 3

8. Answer:

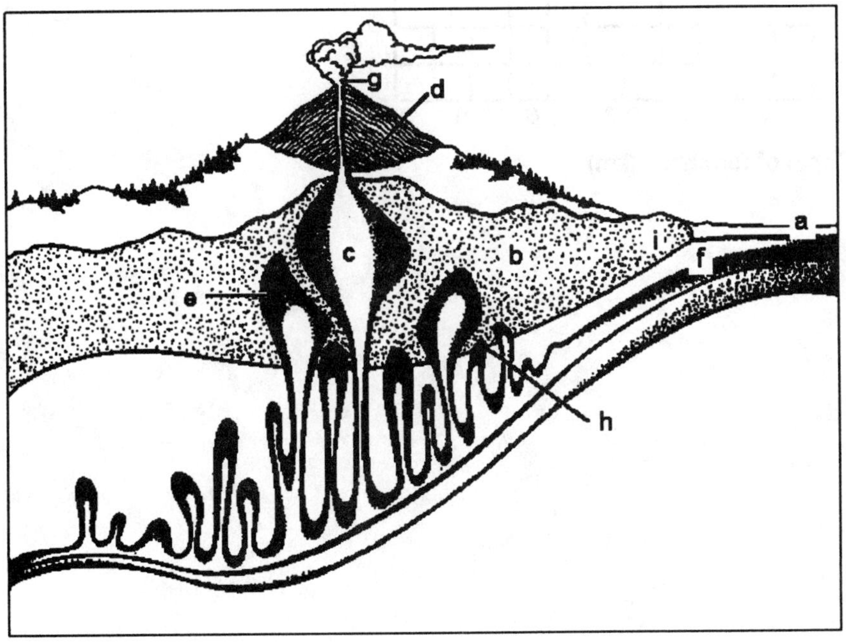

Difficulty: 3

Short Essay

9. Answer:
Sample answer: Relative time describes a series of events based on the order in which the events occurred. For example, the shells were fossilized first and thus are older than the fossilized leaves. A volcano then erupted. The top layer of shale is the youngest of the rock layers.

Difficulty: 4

Graphic

10. Answer:
The smallest grain sizes are found at the shallowest depths. The largest grain sizes are found at the greatest depths because larger particles settle out first.

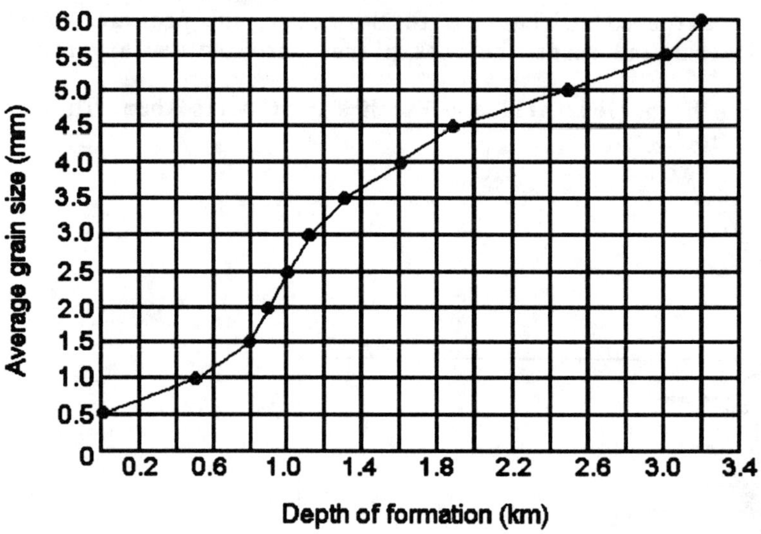

Difficulty: 4

Unit 6 Activity Assessment

Activity Assessment

1. **Modeling Tectonics**

Teacher's Notes

Overview
Students investigate the theory of plate tectonics by constructing a model and re-creating the movement of the Earth's plates.

Materials
(per activity station)
- a shallow rectangular pan, about 30 cm x 40 cm
- water
- a heat source, such as a stove
- oven mitts
- dark food coloring
- 8 pieces of uncorrugated, lightweight cardboard (such as shirt cardboard), 1 cm^2
- a permanent marker
- a metric ruler
- a clock or stopwatch

Preparation
Prior to the assessment, equip student activity stations with the materials needed for each experiment. Remind students to use care when heating the pan.

Time Required
Each student should have 20 minutes at the activity station and 20 minutes to record their responses.

Performance
At the end of the assessment, students should turn in the following:
- a completed Data Sheet

Evaluation
The following is a recommended breakdown for evaluation of this Activity Assessment:
- 30% appropriate use of materials and equipment
- 30% ability to identify the parts of the model as they relate to plate tectonics
- 40% quality and clarity of observations, sketches, and summary

Safety Alert!

2. **Modeling Tectonics**
The theory of plate tectonics helps explain the formation of the Earth's crust and the origins of volcanoes, earthquakes, and mountains. According to the theory, plate movement is caused by currents under the Earth's crust. You can show how this theory works by making a model of the moving plates.

Before You Begin . . .
As you work through the tasks, keep in mind that your teacher will be observing

- how you use the materials and equipment
- how you identify the parts of the model as they relate to the theory of plate tectonics
- the quality of your observations and your understanding of the inaccuracies of the model

Now you're ready to investigate!

What to Do
1. Use the permanent marker to label the cardboard pieces *A-H*.
2. Fill the pan with water up to 3 cm from the top. Place the pan over the heat source, and heat the water for about 30 seconds over a low setting. **Caution! Be careful when using the heat source**.
3. Next, add a few drops of the dark food coloring to the center of the pan. Observe what happens. In the data sheet on the next page, sketch the pattern of movement that you see. Then turn off the heat source.
4. Place cardboard pieces *A*, *B*, *C*, and *D* as close together as possible in the center of the pan. Place one of the remaining pieces in each of the four corners of the pan.
5. Turn the heat source back on. Observe what happens for about 45 seconds, and then sketch the pattern of movement for each cardboard piece. Turn off the heat source when you are finished.
6. Then, in a brief summary, indicate what the water, the cardboard pieces, and the stove represented in your model. Also describe what happened to the food coloring and the cardboard pieces, and relate what you observed to the theory of plate tectonics.
7. Finally, describe why this model isn't completely accurate, and list a suggestion for improvement.
Safety Alert!

3.

Data Sheet

Sketches
Movement of food coloring:
Movement of cardboard pieces:

Summary

____

____

____

____

____

____

____

____

____

____

Suggestion for Improvement

Answers to Unit 6 Activity Assessment

1. Answer: Not applicable (teacher's notes)

2. Answer: Not applicable (student's notes)

3. Answer:

Data Sheet

Sketches
Movement of food coloring: Answers should show that the food coloring moved through the water in a circular pattern.
Movement of cardboard pieces: Answers should show that at first the cardboard pieces in the center of the pan moved apart from one another in a circular pattern. The cardboard pieces in the four corners did not move, however, unless touched by one of the other pieces. Eventually, all of the pieces moved, some in a clockwise circle and others in a counterclockwise circle.

Summary

The heat source in the model represented the Earth's inner core, the water represented part of the Earth that is between the core and the surface, and the cardboard pieces represented the continental plates. The food coloring showed how convection currents created by the heat source move the water in a circular pattern. Those currents can be compared to currents found below the Earth's surface. The cardboard pieces then simulated how currents within the Earth can affect the continental plates by causing them to move, shift, and join together. This is essentially the theory of plate tectonics.

Suggestion for Improvement

Sample answer: The burner on the stove does not extend the full length of the pan, so the heat reaches some of the cardboard pieces before it reaches others. Thus, some pieces begin moving before others. Since heat generated by the Earth's core is more evenly distributed than the model suggests, finding a heat source that could more evenly heat the water would be an improvement.

Unit 6 SourceBook Assessment

Multiple Choice

1. Molten rock is composed of
   a. minerals. b. crystals. c. lava. d. vitamins.

2. _________ rocks are solid masses of crystals.
   a. Sedimentary b. Igneous c. Metamorphic

3. _________ rocks are formed from rock pieces that are broken up by weathering and then stuck together by compaction.
   a. Sedimentary b. Igneous c. Metamorphic

4. _________ rocks are formed when other rocks are exposed to great heat and pressure inside the Earth.
   a. Sedimentary b. Igneous c. Metamorphic

5. Below is an illustration of the rock cycle. Put the appropriate letter in each blank to complete the cycle.
   a. sedimentary rocks b. igneous rocks c. metamorphic rocks

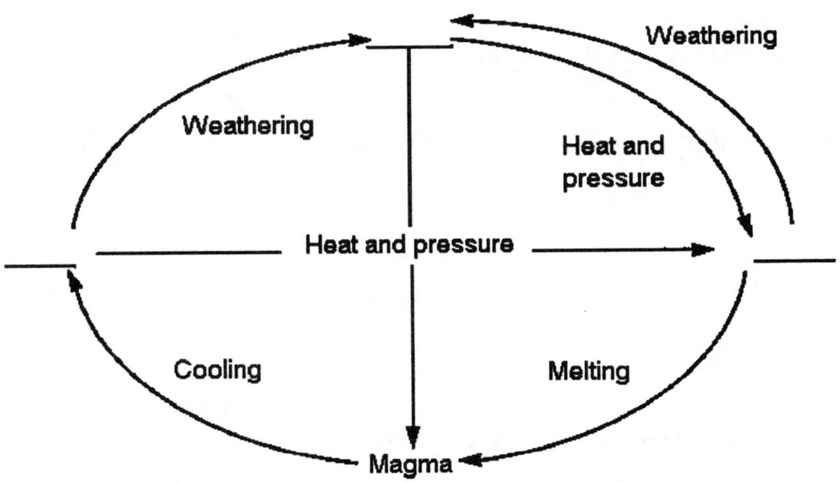

6. Which of the following is not a property used to classify rocks?
   a. color b. texture c. origin d. mineral content

7. ________ igneous rocks are formed deep within the Earth.
   a. Intrusive b. Extrusive

8. Which type of igneous rock takes longer to form?
a. intrusive b. extrusive

9. Rock salt is formed by evaporation and is which type of sedimentary rock?
a. clastic b. chemical c. organic

10. If organisms die and their shells are glued together to form rock, they are this type of rock.
a. igneous
b. clastic sedimentary
c. metamorphic
d. organic sedimentary
e. chemical sedimentary

11. Minerals are divided into silicates and nonsilicates. Which make up the majority of the Earth's crust?
a. silicates b. nonsilicates

12. Some minerals are made of one element; these are called native elements. Which of the following is not an example of a native element?
a. gold b. iron c. calcite d. copper

13. Which of the following is not a property of a mineral?
a. color
b. ability to float in water
c. streak
d. hardness

14. The way a mineral reflects light refers to its
a. color. b. streak. c. reflectivity. d. luster.

15. Younger rock layers lie on top of older rock layers. This demonstrates the principle of
a. unconformity.
b. uniformitarianism.
c. superposition.
d. relativity.

True/False

16. Geologists can determine precisely how many years old a rock is.
a. true b. false

Multiple Choice

17. _______ is the process of estimating the age of a sample of minerals or fossils using radioactive isotopes.
a. Numerical dating
b. Relative age dating
c. Absolute dating
d. both a and c

18. Which type of rock can be dated by the process of numerical dating?
a. sedimentary b. igneous c. metamorphic

True/False

19. Fossils can aid the geologists in determining the age of rocks.
a. true b. false

Multiple Choice

20. Coal is formed from dead plants that were compressed for millions of years. What kind of rock is coal?
a. sedimentary b. igneous c. metamorphic

21. On the mineral hardness scale from 1-10, with 10 being the hardest and 1 being the softest, where would a diamond be?
a. 2 b. 5 c. 7 d. 10

True/False

22. Most earthquakes take place near the boundaries of tectonic plates.
a. true b. false

Multiple Choice

23. The fastest wave created by an earthquake is a
a. transverse wave. b. primary wave. c. secondary wave.

24. A __________ volcano is formed by alternating layers of lava and tephra.
a. shield b. cinder-cone c. composite d. mid-plate

25. Geysers and hot springs are the result of
a. a mid-ocean ridge. b. a hot spot.
c. an earthquake zone. d. a volcano.

Answers to Unit 6 SourceBook Assessment

Multiple Choice

1. Answer: a. minerals.

2. Answer: b. Igneous

3. Answer: a. Sedimentary

4. Answer: c. Metamorphic

5. Answer:

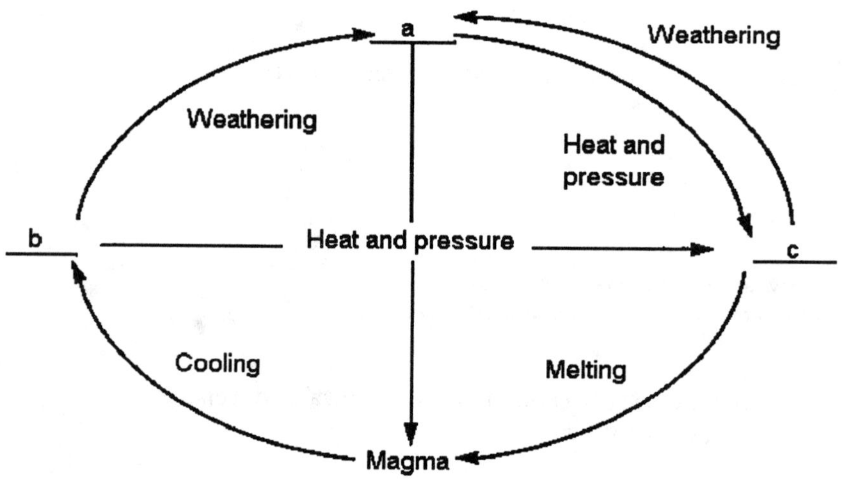

6. Answer: a. color

7. Answer: a. Intrusive

8. Answer: a. intrusive

9. Answer: b. chemical

10. Answer: d. organic sedimentary

11. Answer: a. silicates

12. Answer: c. calcite

13. Answer: b. ability to float in water

14. Answer: d. luster.

15. Answer: c. superposition.

True/False

16. Answer: b. false

Multiple Choice

17. Answer: d. both a and c

18. Answer: b. igneous

True/False

19. Answer: a. true

Multiple Choice

20. Answer: a. sedimentary

21. Answer: d. 10

True/False

22. Answer: a. true

Multiple Choice

23. Answer: b. primary wave.

24. Answer: c. composite

25. Answer: b. a hot spot.

Unit 6 Extra Assessment Items

Word Usage

1. Beside each item, write at least three related words from the list below. Then write a sentence for each item that includes at least one of the related words you chose from the list.

   a. mountain-building: ______________________

   ____

   ____

   b. earthquake: ______________________

   ____

   ____

   c. sedimentary: ______________________

   ____

   ____

   d. metamorphic: ______________________

   ____

   ____

cementation	collide	compaction	continental
convergent	crystallization	divergent	erosion
fault	fold	heat	lava
lithification	Mesozoic	pressure	Richter scale
seismology	sliding	transform	weathering

2. Use the words *Cenozoic* and *era* in one or more sentences that explain the idea of *relative time*.

   ____

   ____

   ____

   ____

Correction/Completion

3. Complete the following statement:
   Metamorphic rocks are rocks that were once sedimentary or igneous but have been changed by . . .

   ____

   ____

Short Essay

4. The crinoid shown below is an extinct species that was once abundant in shallow waters. Modern crinoids attach themselves to the ocean floor with a hard but flexible stalk. They take in food using fanlike appendages that wave around in the moving water. What part of the early crinoid is usually found as a fossil? Why?

   ____

   ____

   ____

   ____

   ____

   ____

   ____

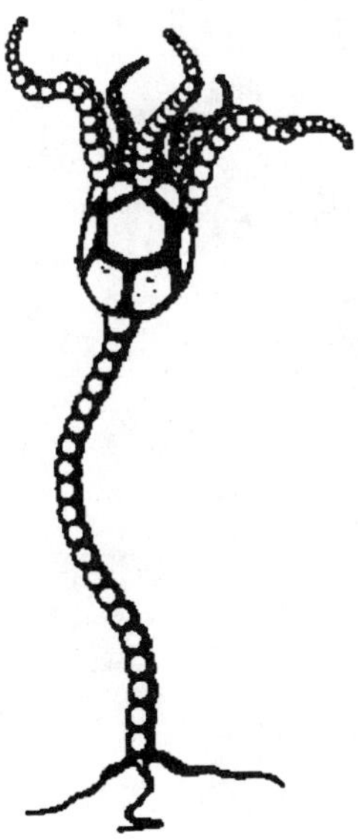

Short Response

5. According to the theory of plate tectonics, what might happen when two plates collide? Use one area of the world as a specific example.

____

____

____

____

6. Identify one part of the world where two plates appear to be moving apart. What events occur between the two plates?

____

____

____

____

____

7. Here are some events in Doug's life.

List A	**List B**
• broke his leg	• 1980—vacationed in Maine
• started high school	• 1991—entered ninth grade
• took a senior trip to Mexico	• 1995—graduated from high school

Which list of events is a relative time scale? Why?

____

____

____

____

Illustrative

8. Norah constructed the graph shown below after collecting data for a geology project.

   a. What relationship, if any, does the graph show between the grain size of the crystals and the depth at which they were formed?

   ____

   ____

   ____

   b. Would you agree with the results shown on Norah's graph, based on what you have studied in this unit? Explain your answer.

   ____

   ____

   ____

   ____

   ____

   ____

   ____

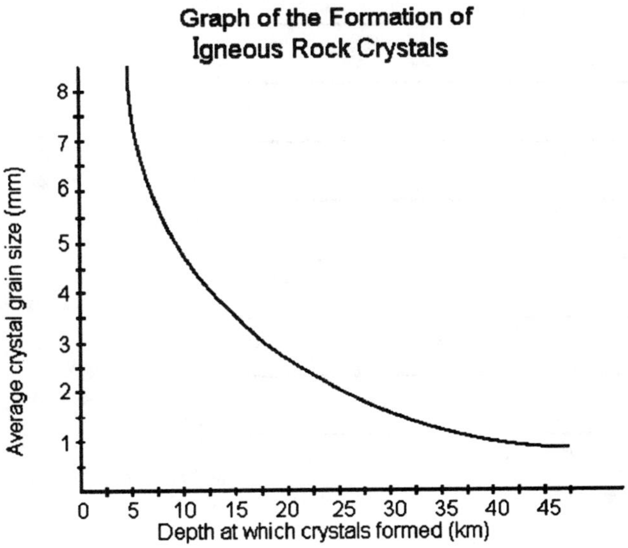

9. You have applied for the job of chief geologist with the state geological survey. As part of your interview for the job, you are asked to review the following illustrations and then answer the questions.

a. Identify the events that led to the scene shown in illustration I.

____

____

b. State what evidence you find in illustration I to support your conclusion.

____

____

____

____

c. Identify the events that led to the scene shown in illustration II.

____

____

d. State what evidence you find in illustration II to support your conclusion.

____

____

____

____

____

____

____

____

____

____

I.

II.

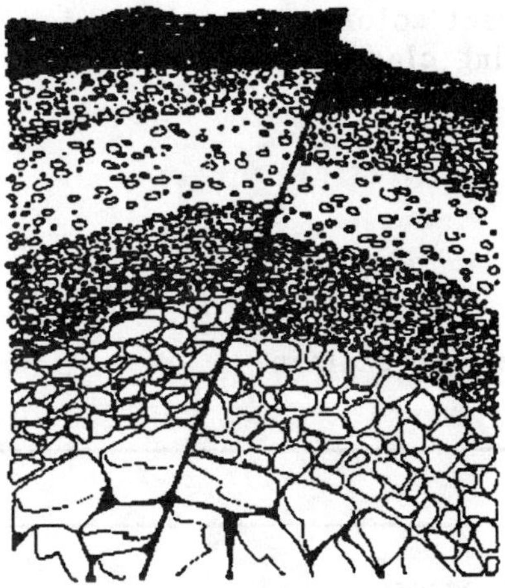

10. Draw a picture showing three ways that people use rocks.

Numerical Problem

11. Use the data chart below to answer the following questions.

Era	Began
Cenozoic	65 million years ago
Mesozoic	190 million years ago
Paleozoic	570 million years ago

a. Which era lasted the longest? ______________________

b. When did the Mesozoic era end? ______________________

c. In which era do you live? ______________________

Performance Task

12. **Materials:** two different colors of modeling clay
Task a: Use the modeling clay to demonstrate what can happen when two continental plates collide. Sketch the result.

What does this model help to explain?

____

____

Name an area where this occurs.

____

Task b: Use the modeling clay to demonstrate what can happen when two plates slide along one another. Sketch the result.

What does this model help to explain?

____

____

Name an area where this occurs.

____

____

13. **Materials:** several examples of igneous, sedimentary, and metamorphic rocks
Task a: Separate the rocks into igneous, sedimentary, and metamorphic groups.
Task b: Describe the origin of each group. In doing this, show that you know the differences between these groups.

Igneous: ____

____

____

Sedimentary: ____

____

____

Metamorphic: ____

____

____

Task c: Now describe each rock and explain why you think it belongs to the group you put it in.

Extended Performance Task

14. **Materials:** a local map, a sketch pad, a field guide to rocks, a geological hammer, safety goggles
Task: Your local geological society has asked for your help. They want to undertake the most thorough study ever of the rock formations in your area. You have volunteered to collect five samples, each of a different kind of rock. If there are solid rock outcrops in your area, you should obtain samples of these. Be certain to obtain permission from all landowners before collecting any rock samples.

For each of the five rocks that you collect, perform the following tasks:
a. On a map, locate the precise location of the sample.
b. Sketch the immediate site from which you took the sample.
c. Was the rock loose or did you pry it from a solid rock bed? If the rock was loose, how do you think it might have gotten where you found it?
d. Identify the rock as igneous, sedimentary, or metamorphic.
e. Use a field guide or the help of a geologist to identify the specific type of rock.
f. Display your results.

Safety Alert: *Be very careful when chipping rock fragments from a rock bed, and watch out for snakes or poisonous plants!*

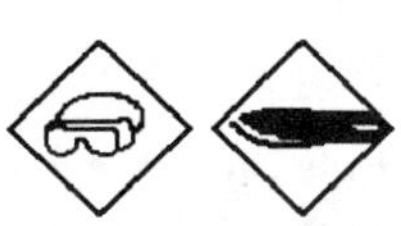

Answers to Unit 6 Extra Assessment Items

Word Usage

1. Answer:
   a. Sample answer: heat, pressure, collide, continental, fold, fault *Mountain-building* occurs when two *continental* plates *collide* and the Earth's crust is pushed upward.
   b. Sample answer: Richter scale, seismology, pressure, collide, fault
   The *Richter scale* measures the intensity of an *earthquake* on a scale of 1 to 10.
   c. Sample answer: pressure, lithification, crystallization, cementation, erosion, weathering, compaction
   Sedimentary rocks are formed by the *lithification* of soft sediments into hard rock.
   d. Sample answer: heat, pressure, transform, fold
   Heat and chemical changes *transform* igneous or sedimentary rock into *metamorphic* rock.

2. Answer:
   Sample answer: The current *era* in geologic history is called the *Cenozoic era*. The Mesozoic era preceded the Cenozoic era.

Correction/Completion

3. Answer:

   Sample answer: ...heat and pressure, which cause chemical changes in the rock.

Short Essay

4. Answer:
   Sample answer: The stalks are found more frequently because they are made up of harder materials. The softer parts are more likely to decay before fossilization can occur. The top parts are also made up of many pieces that scatter easily after death.

Short Response

5. Answer:
   Students will probably mention the occurrence of a fault, earthquake, or volcano, or the creation of a mountain. Specific examples will vary but should be accurate examples of the situation described.

6. Answer:
   Answers should mention the mid-Atlantic ridge or one of the other ocean ridges where the separation of plates occurs. The answer should also describe the events that occur there: volcanoes, earthquakes, and the formation of new crust.

7. Answer:
   Sample answer: List A is a relative time scale. A relative time scale has items arranged in chronological order but without specific dates.

Illustrative

8. Answer:
   a. The largest crystals are found at the shallowest depths; the smallest crystals are found at the greatest depths.
   b. Sample answer: I would disagree with Norah's graph. The more quickly an igneous rock cools, the smaller its crystals are. Molten rock close to the surface cools more rapidly than molten rock found deep underground. Therefore, the closer the molten rock is to the surface, the faster it will cool and the smaller its crystals will be. (At great depths pressure may create smaller crystals, but this should not be an issue for the depths shown on the graph.)

9. Answer:
   a. It appears that a major fault cut across the river.
   b. The displacement of the upper part of the river from the lower part along a fault line is evidence that movement along the fault probably occurred after the river was formed.
   c. A major fault evidently occurred after a somewhat complex geological history.
   d. The probable sequence of events that led to the pattern shown in the illustration is as follows: 1) The bottom layers that are now folded and cut were probably deposited originally as flat layers, perhaps as sediments on the bottom of a body of water. 2) Later, folding occurred, probably associated with continental drift and mountain-building. 3) Erosion created a flat surface. 4) Further layers were deposited on top of that surface, probably when it was again at the bottom of a body of water. 5) Major faulting then occurred, causing the left-hand side to be lifted over the right-hand side. 6) Subsequently, further surface erosion occurred, giving the surface its contours.

10. Answer:
    Answers will vary but should be logical and complete. Answers could include walls, homes, pavement, decoration, etc.

Numerical Problem

11. Answer:
    a. Paleozioc
    b. 65 million years ago
    c. Cenozoic

Performance Task

12. Answer:
Task a: Sketches should show a portion of the Earth's crust that has buckled and moved upward.

This model helps to show how the Earth's crust can be moved upward to create hills and mountains.

Answers will vary, but most large hills or mountains, such as the Himalayan mountains, would qualify.

Task b: Sketches should show two blocks of modeling clay positioned parallel to one another, with one block in front of the other.

This model helps to explain a transform fault that causes earthquakes in California.

Answers will vary but should name a fault line or an area where earthquakes have previously occurred.

13. Answer:
Task b:
Igneous: These rocks originate as magma cools and solidifies either on or below the Earth's surface.

Sedimentary: These rocks originate when rock particles are deposited by wind, water, ice, or chemical reactions.

Metamorphic: These rocks originate when any rocks undergo change as a result of intense heat and pressure.

Task c: Answers will vary depending on the specific rocks chosen.

14. Answer:
Answers will vary depending on the specific circumstances, but students should collect five rocks and correctly complete the above tasks for each rock.

Chapter 19 Assessment

Word Usage

1. a. Create at least three astronomical terms using the letters in the word *astronomy*. You may choose one optional letter that is not in the word *astronomy* for each word you create.

   ____

   ____

   ____

   b. Use the following terms in one or two sentences to show how they are related: Sun, shadow, time, sundial.

   ____

   ____

   ____

Correction/Completion

2. The statements below are either incorrect or incomplete. Your challenge is to make them correct and complete.

   a. The winter sky looks exactly like the summer sky.

   ____

   ____

   ____

   b. Aristotle was the first scientist to measure the distances from the Earth to the Sun and the Moon.

   ____

   ____

   ____

Short Response

3. The Flat Earth Society is a group of people who claim to still believe that the Earth is flat. What are two pieces of evidence that you can give to prove that the Earth is round?

____

____

____

____

____

____

____

4. Design an experiment to tell whether the same star will appear earlier or later on one night than it will on the next night.

____

____

____

____

____

____

Illustrative

5. Most early astronomers had a view of the solar system that is different from the view of modern scientists. Make two drawings to show how these views are different.

Short Essay

6. Why was it important for early civilizations to track time?

____

____

____

____

____

Answers to Chapter 19 Assessment

Word Usage

1. Answer:
   a. Possible terms include star, Moon, Mars, Sun, Saturn, solar, and NASA.
   b. Sample answer: A *sundial* is an ancient tool for telling *time* by measuring the *shadow* of the *Sun*.

Correction/Completion

2. Answer:
   a. Sample answer: The winter sky looks *different from* the summer sky *because different constellations are visible during particular seasons.*
   b. Sample answer: *Aristarchus of Samos* was the first scientist to measure the distances from the Earth to the Sun and the Moon.

Short Response

3. Answer:
   Student answers may include evidence such as the following: photographs of Earth taken from space; the curved horizon; viewing other planets with telescopes.

4. Answer:
   Sample answer: Observe a star for at least two consecutive nights. Record the exact time it appears on the first night. Begin the observation several minutes earlier the second night. Note when the star appears and compare the times. (The same star will be seen in the same position about 4 minutes earlier on the second night.)

Illustrative

5. Answer: Sample drawings:

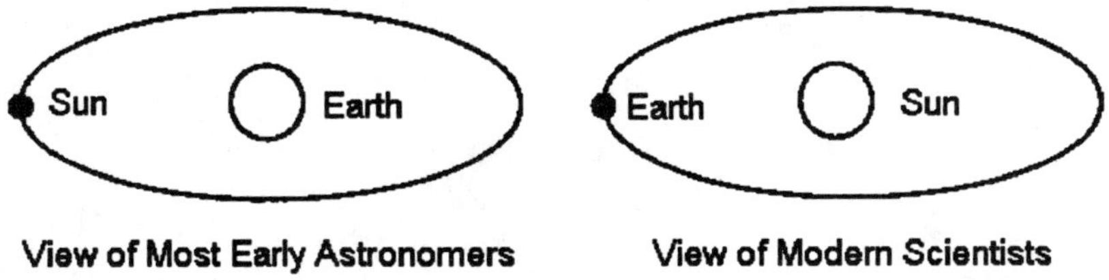

Difficulty: 3

Short Essay

6. Answer:
   Sample answer: Early civilizations developed sundials and calendars so that they would know when to plant and harvest their crops. These timekeeping devices provided information about important factors such as the amount of daylight and changing seasons.

   Difficulty: 4

Chapter 20 Assessment

Word Usage

1. a. People frequently confuse the terms *rotation* and *revolution*. Use them in one or two sentences to show their correct usage.

____

____

____

____

b. Use the words *Sun* and *Earth* to define the term *plane of the ecliptic*.

____

____

____

Short Response

2. Galileo and Ptolemy had different ideas about the organization of the solar system. Write the letter *G* by the statements that support Galileo's idea and the letter *P* by the statements that support Ptolemy's idea.

___ The Sun is the center of the solar system.

___ Mars moves in small circles as it orbits the Earth.

___ If the Earth really moved, tremendous winds would result. These winds would blow down trees.

___ Night and day are caused by the rotation of the Earth as it revolves around the Sun.

___ The Earth is stationary and the constellations orbit the Earth.

___ The apparent backward motion of Mars is an optical illusion caused by the different rates at which Mars and Earth revolve around the Sun.

Numerical Problem

3. The average distance from the Earth to the Sun is 150 million kilometers (1 AU). If light travels at 300,000 km per second, how long does it take for sunlight to reach Earth? Show your work.

____

____

Illustrative

4. Use the illustration to determine which city or cities would receive the most sunlight on each date listed below.

June 21 ____________

March 21 ____________

December 21 ____________

September 21 ____________

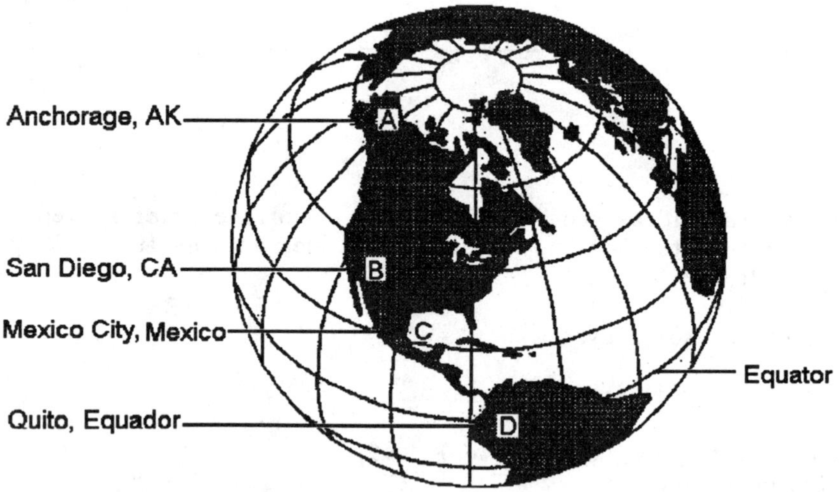

5. The following observation of the Sun was made by a science class over a period of time. Draw a labeled diagram to explain the observation.

• Every day, the Sun rose in the east and set in the west.

Short Essay

6. Suppose that the Earth were **not** tilted on its axis. How would our days, nights, seasons, and weather be affected?

Answers to Chapter 20 Assessment

Word Usage

1. Answer:
   a. Sample answer: *Rotation* of the Earth on its axis causes day and night. The *revolution* of the Earth around the Sun causes the seasons.
   b. The *plane of the ecliptic* is the plane in which *Earth* and the other planets orbit the *Sun*.

Short Response

2. Answer:
   G The Sun is the center of the solar system.
   P Mars moves in small circles as it orbits the Earth.
   P If the Earth really moved, tremendous winds would result. These winds would blow down trees.
   G Night and day are caused by the rotation of the Earth as it revolves around the Sun.
   P The Earth is stationary and the constellations orbit the Earth.
   G The apparent backward motion of Mars is an optical illusion caused by the different rates at which Mars and Earth revolve around the Sun.

Numerical Problem

3. Answer:

$$\frac{150 \text{ million km}}{300{,}000 \text{ km per sec.}} = 500 \text{ sec.}$$

$$\frac{500 \text{ sec.}}{60 \text{ sec. per min.}} = 8.3 \text{ min.}$$

Illustrative

4. Answer:

June 21	A
March 21	A, B, C, D
December 21	D
September 21	A, B, C, D

5. Answer: Sample answer:

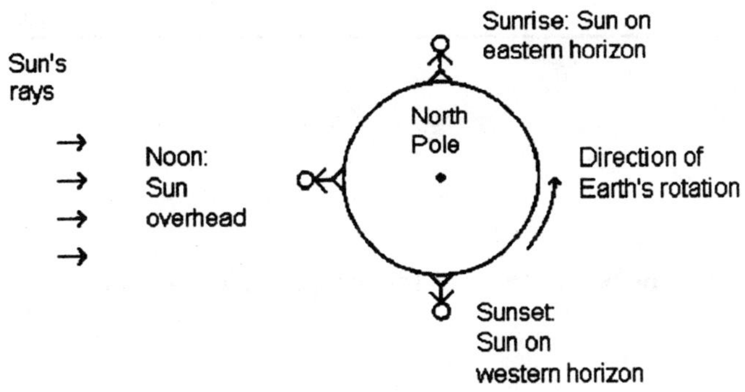

Difficulty: 3

Short Essay

6. Answer:
Sample answer: As long as the Earth still rotated on its axis, we would still have days and nights. However, the amount of daylight and darkness would always be equal, and we would not have seasons because the North and South Poles would never be tilted away from the direct rays of the Sun. Therefore, our weather would remain more constant throughout the year.

Difficulty: 4

Chapter 21 Assessment

Word Usage

1. Write a sentence to explain the difference between the following pairs of words:

   a. meteor, meteorite

   b. comet, asteroid

Correction/Completion

2. The statements below are either incorrect or incomplete. Your challenge is to make them correct and complete.

   a. Shooting stars are stars that orbit the Sun at a rapid speed.

   b. The satellite that took pictures of Venus is called *Voyager 1*.

Short Response

3. Put an "X" by the objects that orbit the Sun.

___ asteroids ___ comets

___ meteorites ___ planets

___ stars ___ *Voyager 1*

Illustrative

4. Consider the diagram below and answer the questions that follow.

   a. Name two celestial bodies found in the solar system that are not pictured here.

   ____

   ____

   b. Which planets could you call the "gaseous giants"? Why is this a good name for them?

   ____

   ____

   ____

   ____

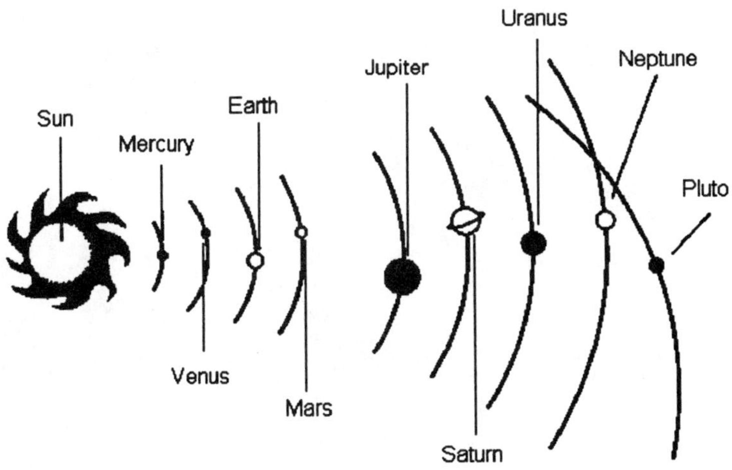

Short Essay

5. Imagine that you are on a voyage of discovery to visit two of the planets in the solar system. Describe your trip. In your description, tell about the appearance of each of these planets as you approach it. Could you land on the planet safely? Why or why not?

Answers to Chapter 21 Assessment

Word Usage

1. Answer:
   a. Sample answer: A *meteorite* is a *meteor* that strikes the Earth's surface rather than burning up in the atmosphere.
   b. Sample answer: A *comet* and an *asteroid* are both elements of the solar system, but asteroids are large rocks or boulders, while comets are made of ice and have tails when they are close to the Sun.

Correction/Completion

2. Answer:
   a. Shooting stars are *rock particles that have entered the Earth's atmosphere during a meteor shower.*
   b. The satellite that took pictures of Venus is called *Magellan.*

Short Response

3. Answer:

X asteroids X comets
___ meteorites X planets
___ stars ___ *Voyager 1*

Illustrative

4. Answer:
   a. Answers may include moons, comets, meteors, asteroids, or artificial satellites.
   b. Jupiter, Saturn, Uranus, and Neptune are the gaseous giants. This is a good name for them because, compared to the other planets, they are much larger and are made up primarily of gases instead of rock.

   Difficulty: 3

Short Essay

5. Answer:
   Many answers are possible depending on the two planets that are visited. Check the answer that is given against the photographs and descriptions in Lessons 2 and 3 of your textbook. Consider the answer acceptable if two correct pieces of information are provided about each planet visited.

   Difficulty: 4

Chapter 22 Assessment

Word Usage

1. a. Using the words *hydrogen*, *helium*, and *nuclear reactions*, explain how we feel the heat and see the light from stars.

   ____

   ____

   ____

   ____

   b. Explain the history and present state of Betelgeuse using the words *red supergiant*, *fuel*, *diameter*, and *supergiant*.

   ____

   ____

   ____

   ____

Correction/Completion

2. The statements below are either incorrect or incomplete. Your challenge is to make them correct and complete.

   a. A light-year is a unit of time.

   ____

   b. The Sun is a medium-sized __________ in the __________ __________ galaxy.

   c. A star's color and apparent brightness depend on its __________,

   __________, and __________ from Earth.

Short Response

3. a. Number the following stars and their colors from coolest to hottest.

___ the Sun, yellow
___ Betelgeuse, red
___ Sirius, blue

b. Number the following list of objects from least dense (1) to most dense (4).
___ neutron star
___ black hole
___ red giant
___ supernova

Numerical Problem

4. a. How many kilometers are in a light-year? Show your work.

____

____

____

____

____

b. The star Alpha Centauri is 4.3 light-years from Earth. Suppose that it explodes on January 1, 2005. When would we see the explosion on Earth?

____

____

____

Short Essay

5. Explain the big-bang theory in your own words.

Answers to Chapter 22 Assessment

Word Usage

1. Answer:
   a. Sample answer: Inside a star's core, *nuclear reactions* occur in which *hydrogen* atoms fuse together to form *helium*. During the reaction, energy is released in the form of heat and light.
   b. Sample answer: Ten million years ago, Betelgeuse was a *supergiant*, but after burning most of its hydrogen *fuel*, its *diameter* has expanded to 700 times that of the Sun. It is now a *red supergiant*.

Correction/Completion

2. Answer:
   a. A light-year is a unit of *distance*.
   b. The Sun is a medium-sized <u>star</u> in the <u>Milky</u> <u>Way</u> galaxy.
   c. A star's color and apparent brightness depend on its <u>size</u>, <u>temperature</u>, and <u>distance</u> from Earth.

Short Response

3. Answer:
   a. <u>2</u> the Sun, yellow
      <u>1</u> Betelgeuse, red
      <u>3</u> Sirius, blue
   b. <u>3</u> neutron star
      <u>4</u> black hole
      <u>1</u> red giant
      <u>2</u> supernova

Numerical Problem

4. Answer:
   a. Light travels at a speed of 300,000 km/s

$$300{,}000 \text{ km/s} \times \frac{60 \text{ s}}{1 \text{ minute}} \times \frac{60 \text{ min.}}{1 \text{ hour}} \times \frac{24 \text{ hr.}}{1 \text{ day}} \times \frac{365 \text{ days}}{1 \text{ year}} =$$

   9,460,800,000,000 km

   b. We would see the explosion 4.3 years later, sometime around April 15, 2009.

   Difficulty: 3

Short Essay

5. Answer:
   Sample answer: According to the big-bang theory, the universe as we know it began 10 to 20 billion years ago with a cosmic event, and it has been expanding ever since. About 14 billion years ago, as the universe cooled, galaxies formed. The objects called quasars that we can detect today with giant radio telescopes may be galaxies in the early stages of formation. About 5 billion years ago, the Sun and the planets in our solar system formed.

   Difficulty: 4

Unit 7 End-of-Unit Assessment

Word Usage

1. a. Write a sentence using the following words to show how they are related: astronomy, science, stars, universe.

____

____

____

   b. Use the words *constellation*, *stars*, and *Orion* in a sentence to show your understanding of what a constellation is.

____

____

____

   c. Use the words *galaxies*, *universe*, *big bang*, and *cosmic expansion* to describe one theory of how the universe was formed.

____

____

____

____

2. Use the words below to complete the diagram, which shows the organization of the universe.

Earth, Orion Arm, Pluto, Sun, Saturn, Milky Way galaxy, Local Group

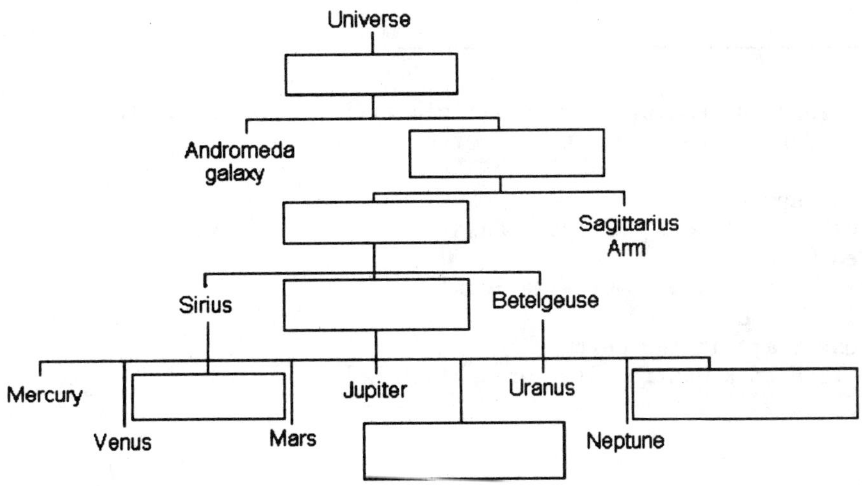

Short Response

3. List at least six different objects in our solar system.

4. Draw a line to match the following speakers with their ideas about astronomy.

Aristotle	• If the Earth really moved, the tremendous winds would blow birds from the sky and knock down trees.
Galileo	• The orbit of each planet is shaped as an ellipse—not a circle!
Kepler	• Like Aristarchus of Samos, I believe that Earth and all the other planets revolve around the Sun.
Ptolemy	• . . . And still the Earth moves.
Copernicus	• Everything in the universe consists of four elements: earth, air, fire, and water.

5. Address a letter to yourself using your cosmic address. Be sure to include your galaxy!

6. Here are a number of paragraph headings for an article titled "Life of a Star." They are not in the proper order. Number them in the correct order.

___ a. Star Explodes as Supernova
___ b. Millions, or Perhaps Billions, of Years Pass
___ c. Star Forms a Red Giant
___ d. Hydrogen Clumps Together Due to Gravitational Force
___ e. Hydrogen Fuel Is Used Up
___ f. Nuclear Reactions Start in the Core
___ g. Star Collapses to Form a Neutron Star or a Black Hole

Data for Interpretation

7. Consider the information in the table below. Then answer the questions that follow.

Planet	Average distance (in AU)	Time it takes to revolve once around the Sun (in Earth days)
Mercury	0.4	88
Venus	0.7	225
Earth	1.0	365
Mars	1.5	687

a. One astronomical unit is equal to the distance between what two objects?

____

____

____

b. Which planet is closest to Earth?

____

____

c. Imagine that years are now measured by the time it takes Venus to revolve around the Sun. What is your age in Venus years?

____

____

Illustrative

8. The diagram below has a number of errors in it. Show what corrections need to be made to the diagram.

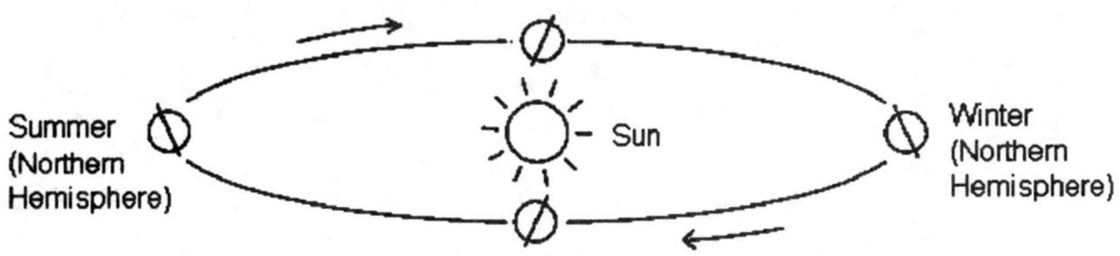

9. The diagram on the left shows the position of Polaris (the North Star) and the Big Dipper at 10:00 P.M. In the space on the right, draw the position of these celestial objects at midnight.

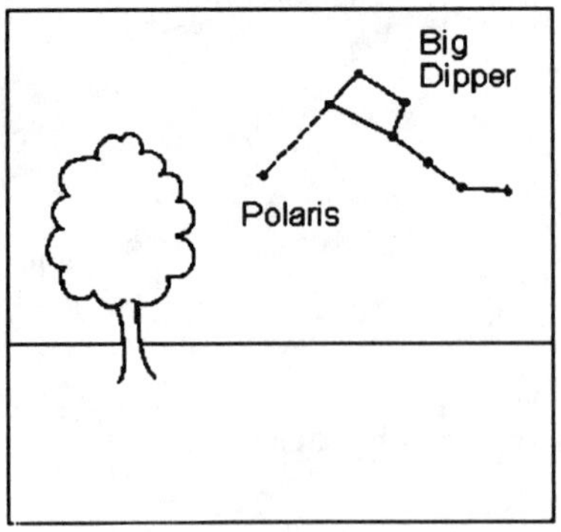

10:00 p.m.

Midnight

10. Draw a diagram of how the Sun's rays strike the Earth in order to illustrate each of the following conditions. Then tell what seasonal event is being described.
a. The number of hours of daylight equals the number of hours of darkness.

b. The number of hours of daylight (in the Northern Hemisphere) is greatest.

____

Short Essay

11. Be a star! Write a brief paragraph about the life of a star as if you were the star. Include the following elements:
    - how you were formed
    - the stages you passed through
    - your eventual fate

Numerical Problem

12. After leaving Earth and traveling for approximately 16 million years at light speed, you are able to see the entire Local Group of galaxies. To reach the edge of the known universe, you must travel at least 1000 times farther. Answer the questions that follow about the size of the universe. (You may round off your figures.)

    a. How many light-years would it take to reach the edge of the known universe?

    b. Knowing that 1 light-year = 63,000 AU, how far would you have to travel in astronomical units?

    c. If 1 AU = 150,000,000 km, how far would you have traveled in kilometers?

Answers to Unit 7 End-of-Unit Assessment

Word Usage

1. Answer:
   a. Sample answer: The *science* of studying the *stars* and other bodies in the *universe* is called *astronomy*.
   b. Sample answer: *Orion* is a *constellation*, or pattern of *stars*, that is seen in the winter sky.
   c. Sample answer: The *universe* probably began from a great *cosmic expansion* called the *big bang*. This may explain why *galaxies* are moving away from each other.

2. Answer:

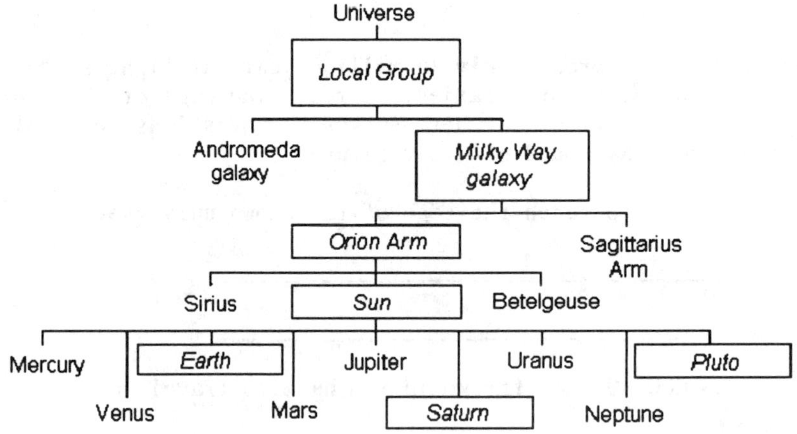

Short Response

3. Answer:
   Possible answers include the Sun, the Moon, the nine planets, the asteroid belt, comets, meteors, and artificial satellites.

4. Answer:

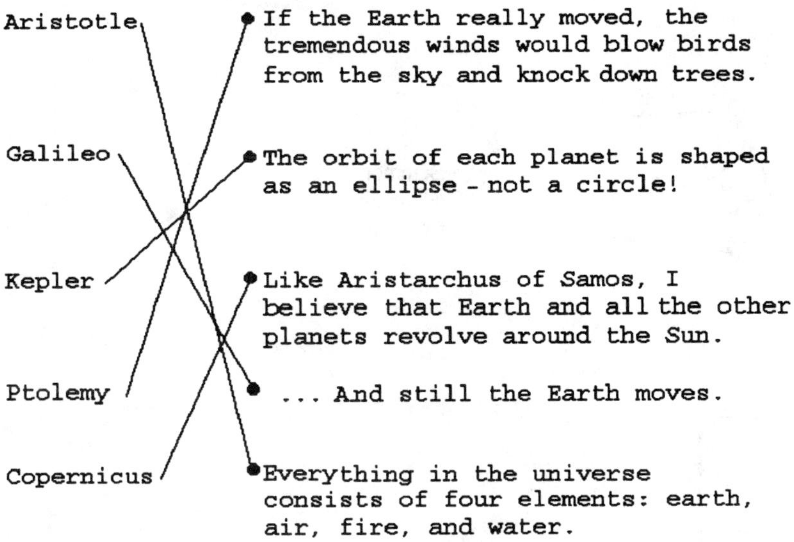

5. Answer:

Name
Street
City
State
United States
North America
Northern Hemisphere
Earth
Solar System
Orion Arm
Milky Way
Local Group

6. Answer:

6 a. Star Explodes as Supernova
3 b. Millions, or Perhaps Billions, of Years Pass
5 c. Star Forms a Red Giant
1 d. Hydrogen Clumps Together Due to Gravitational Force
4 e. Hydrogen Fuel Is Used Up
2 f. Nuclear Reactions Start in the Core
7 g. Star Collapses to Form a Neutron Star or a Black Hole

Data for Interpretation

7. Answer:

a. One astronomical unit is equal to the distance between the Earth and the Sun.
b. Venus is closest to Earth.

c. Sample answer for a 14-year-old student: $\frac{365 \text{ days}}{225 \text{ days}} \times 14 = 22.7$ Venus years

Illustrative

8. Answer:

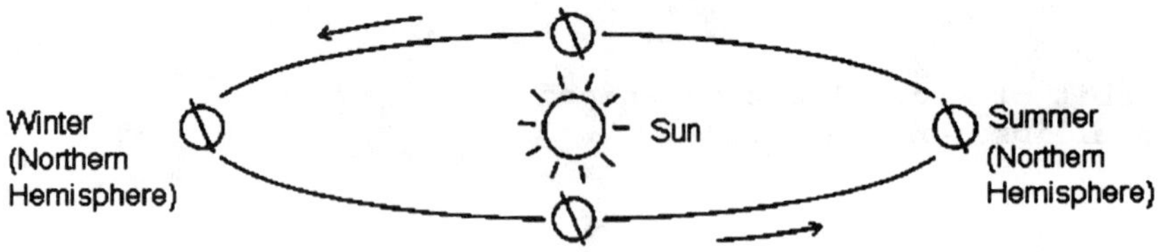

9. Answer:

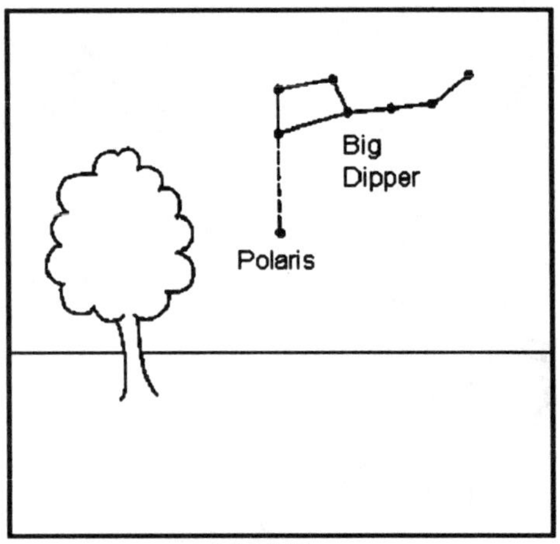

Midnight

Difficulty: 3

10. Answer: Sample answers:

a.

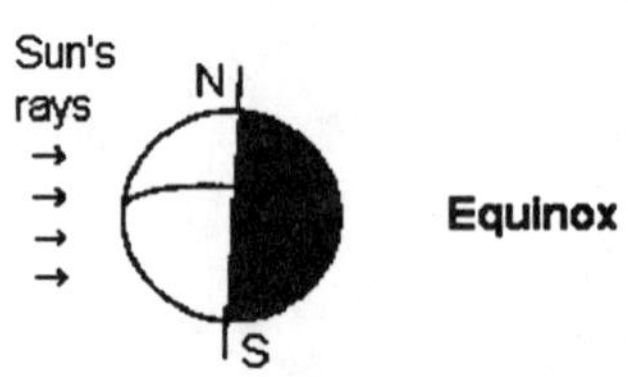

Either the vernal or autumnal equinox.

b.

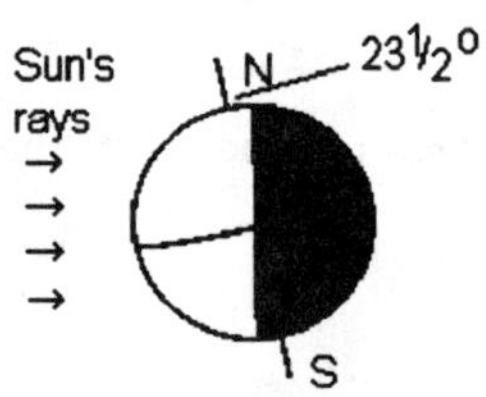

The summer solstice

Difficulty: 3

Short Essay

11. Answer:

Sample answer: Hydrogen groups together due to gravitational force; nuclear reactions start at the core; millions or billions of years pass; the hydrogen fuel is used up; the star forms a red supergiant; the star explodes; the star collapses to form a neutron star or a black hole.

Difficulty: 4

Numerical Problem

12. Answer:

a. 16,000,000 light-years x 1000 = 16 billion light-years

b. $16{,}000{,}000{,}000 \text{ light-years} \times \frac{63{,}000 \text{ AU}}{1 \text{ light-year}} = 1.0 \times 10^{15} \text{ AU}$

c. $1.0 \times 10^{15} \text{ AU} \times \frac{150{,}000{,}000 \text{ km}}{1 \text{ AU}} = 1.5 \times 10^{23} \text{ km}$

Difficulty: 4

Unit 7 Activity Assessment

Activity Assessment

1. **Phases of the Moon**

Teacher's Notes

Overview
Students use a model of the Moon to demonstrate their understanding of the different phases of the Moon. They sketch three different phases and draw conclusions about why the Moon sometimes appears as a crescent.

Materials
(per activity station)
- a globe of the Earth
- a plastic-foam ball
- a light source

Preparation
Prior to the assessment, equip student activity stations with the materials needed for each experiment. The activity should be performed in a darkened room.

Time Required
Each student should have 25 minutes at the activity station and 15 minutes to complete their sketches.

Performance
At the end of the assessment, students should turn in the following:
- a completed Observation Sheet

Evaluation
The following is a recommended breakdown for evaluation of this Activity Assessment:
- 20% appropriate use of materials and equipment
- 30% ability to simulate the phases of the Moon
- 30% quality and clarity of observations, sketches, and summary
- 20% ability to draw conclusions based on observations

2. **Phases of the Moon**
It's easy to tell when the Moon is full. But you may have wondered exactly what you see when the Moon appears as a crescent or when you cannot see the Moon at all. Does the Earth cast its shadow on the Moon? Is all of the moonlight reflected from the Sun? At your activity stations you will find the materials you need to model different phases of the Moon. Draw your own conclusions about how and why the Moon appears as it does in each phase.

Before You Begin . . .
As you work through the tasks, keep in mind that your teacher will be observing:
- how you use the materials and equipment
- how well you simulate the different phases of the Moon
- the clarity of your sketches
- how you draw conclusions about the different phases of the Moon

Now you're ready to light up the world!

Task 1 Locate your state on the globe. Using the light source as your model for the Sun, adjust the globe so that your area is in the middle of summer.

Task 2 Using the ball as your model for the Moon, pass the Moon between the Earth and the Sun. The side of the Moon that faces the Earth will be in darkness. On your Observation Sheet, make a sketch of this new-Moon phase. Show the positions of the Earth, Moon, and Sun.

Task 3 Continue to move the Moon in its orbit around the Earth. When part of the model Moon is illuminated by the light source, the Moon is in the crescent phase. Sketch this phase as it would appear in relation to the new-Moon phase.

Task 4 About 2 weeks after the new Moon, the entire Moon is visible in the sky. Using your materials, model this event. Then add this full-Moon phase to your diagram.

Task 5 What other phases can your add to your diagram? How about the quarter Moon? the gibbous Moon?

Task 6 Draw your own conclusions about why the Moon sometimes appears as a crescent to viewers on Earth.

3. **Observation Sheet**

Sketches:

Conclusions:

Answers to Unit 7 Activity Assessment

1. Answer: Not applicable (teacher's notes)

2. Answer: Not applicable (student's notes)

3. Answer:
   Sample answer: The Moon revolves around the Earth and half of the Moon's surface is always lit. Thus, there are positions where, because of its angle relative to Earth, only a portion of the sunlit side of the Moon can be seen from Earth. This is when the Moon appears as a crescent to viewers on Earth.

   Sample student illustration:

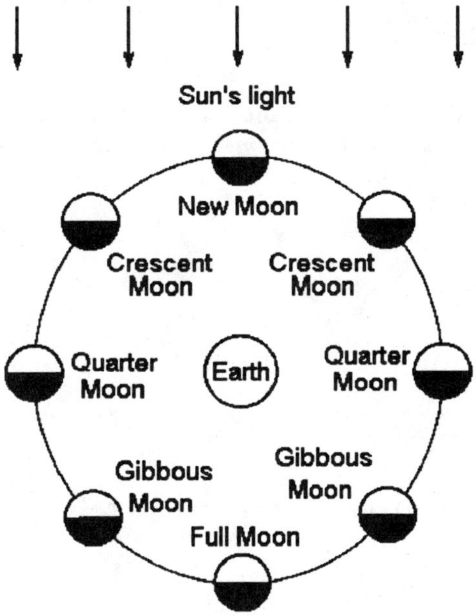

Unit 7 SourceBook Assessment

Multiple Choice

1. The Moon (does/does not) have an atmosphere.

2. The most important source of energy used on Earth is
   a. electricity. b. the Sun. c. the wind. d. geothermal.

3. When we see the Sun's image, we are looking at the glowing layer called the
   a. photosphere. b. corona. c. core. d. chromosphere.

4. Solar wind is part of which layer of the Sun?
   a. photosphere b. corona c. core d. chromosphere

5. If you were in a spaceship traveling away from the Sun, which planet would you reach last? (Assume that we have the technology for such a space flight and that the planets are in alignment with each other.)
   a. Jupiter
   b. Saturn
   c. Pluto
   d. Uranus
   e. None of the above

6. Fragments of rock that orbit the Sun are called
   a. satellites. b. meteors. c. asteroids. d. comets.

7. Astronomers use _______ and _______ to estimate the distance to stars. (Choose two answers.)
   a. luminosity
   b. absolute brightness
   c. parallax
   d. theory of relativity

8. From a star's light we can learn
   a. what it is made of.
   b. how it is moving in space.
   c. how big it is.
   d. All of the above

9. Many of the stars we see at night are actually two stars that revolve around each other. These are called
   a. twin stars. b. binary stars. c. duets. d. satellite pairs.

10. The Milky Way galaxy is what shape?
    a. spiral b. barred spiral c. elliptical d. irregular

11. A theoretical object so dense that its gravity will allow nothing to escape from it is called a
    a. quasar. b. pulsar. c. variable. d. black hole.

12. Telescopes pick up
    a. visible light.
    b. radio waves.
    c. gamma rays.
    d. All of the above

13. Which space mission landed on the moon in 1969, allowing Neil Armstrong and Edwin Aldrin to be the first humans to walk on the Moon?
    a. *Gemini 5* b. *Sputnik 2* c. *Explorer 1* d. *Apollo 11*

14. If you want to learn about how the universe began and how it may eventually end, you should study
    a. cosmology.
    b. astrology.
    c. cosmetology.
    d. planetary science.

15. The space probe that provided information about the Sun was
    a. *Voyager II*.
    b. *Magellan*.
    c. *Ulysses*.
    d. *Pioneer 10*.

16. The northern lights (aurora borealis) are caused by
    a. meteors.
    b. comets.
    c. particles from the Sun.
    d. asteroids.

True/False

17. There is water on the surface of the Moon.
    a. true b. false

18. Most of the outer planets are made of ices and gases, and most of the inner planets are made of solid rock and metals.
    a. true b. false

19. The higher the temperature of a star is, the brighter the star is.
    a. true b. false

20. Our Sun is located in the middle of the Milky Way galaxy.
    a. true b. false

21. Reflecting telescopes use lenses, not mirrors, to view distant objects.
    a. true b. false

Short Response

22. Knowing that the color of a star's light is related to the temperature of the star, match the star type on the left with the appropriate surface temperature on the right.

a. red stars ________ hot (10,000°C)

b. blue stars ________ warm (6000°C)

c. yellow stars ________ cool (3000°C)

23. Number each of the following according to size, with 1 being the smallest and 6 being the largest.

________ galaxy ________ Mars ________ Sun

________ Earth ________ Saturn ________ Venus

24. The light coming from distant galaxies is shifted to the red end of the spectrum. What does this tell us about the galaxies?

____

____

____

____

25. Use what you have learned in this SourceBook unit to briefly explain the big-bang theory.

____

____

____

____

____

____

Answers to Unit 7 SourceBook Assessment

Multiple Choice

1. Answer: does not

2. Answer: b. the Sun.

3. Answer: a. photosphere.

4. Answer: b. corona

5. Answer: c. Pluto

6. Answer: c. asteroids.

7. Answer:
   a. luminosity
   c. parallax

8. Answer: d. All of the above

9. Answer: b. binary stars.

10. Answer: a. spiral

11. Answer: d. black hole.

12. Answer: d. All of the above

13. Answer: d. *Apollo 11*

14. Answer: a. cosmology.

15. Answer: c. *Ulysses*.

16. Answer: c. particles from the Sun.

True/False

17. Answer: b. false

18. Answer: a. true

19. Answer: a. true

20. Answer: b. false

21. Answer: b. false

Short Response

22. Answer:

a. red stars	b	hot (10,000°C)
b. blue stars	c	warm (6000°C)
c. yellow stars	a	cool (3000°C)

23. Answer:

6 galaxy 1 Mars 5 Sun
3 Earth 4 Saturn 2 Venus

24. Answer:
An apparent red shift in the light coming from distant galaxies indicates that the galaxies are moving away from each other.

25. Answer:
Answers will vary, but the following ideas should be expressed: All matter and energy were contained in one very small, dense sphere. Then there was a tremendous event that caused the universe to begin expanding, sending matter outward in all directions. This matter was then pulled together by gravity, forming all of the galaxies, planets, stars, etc.

Unit 7 Extra Assessment Items

Word Usage

1. Use the words *expanding* and *big bang* in one or more sentences with the word *universe* to describe a theory of how the universe began.

2. Writers frequently confuse the terms *planet* and *satellite*. Use the terms in a sentence to show their correct usage.

3. Use the words *crater*, *meteorite*, and *meteor* in an explanation of how craters on Earth might have been formed.

Short Essay

4. *NEWS FLASH!! Astronomers have discovered a new energy source 3 billion light-years from Earth. The astronomers believe that they are seeing a galaxy during its early stages of formation 3 billion years ago.*
   Continue this newspaper article by informing your readers of the meaning of light-years and explaining why astronomers are seeing this energy source as it appeared in the past.

5. Imagine that you are a science writer for a newspaper. Write a newspaper-style article for one of the following headlines. Be sure to give your readers scientific information about the subject.

   Amateur Astronomer Discovers New Comet
   Supernova Observed in Andromeda Galaxy
   Near Collision Between Asteroid and Space Probe

6. Suppose that you are able to travel through the universe. Describe two of the components of the universe that you might encounter beyond our solar system. (You may use sketches to support your descriptions, if you wish.)

____

____

____

____

____

7. a. Write five questions that a person might be able to answer after studying astronomy.

____

____

____

____

____

____

____

____

b. Answer two of these questions yourself, based on what you have learned from studying astronomy.

____

____

____

____

____

____

____

____

Short Response

8. What is the big-bang theory? According to this theory, has our universe always existed as we see it today? If not, how is it changing?

____

____

____

____

____

____

____

Correction/Completion

9. Complete the diagram below, which illustrates the life of a star.

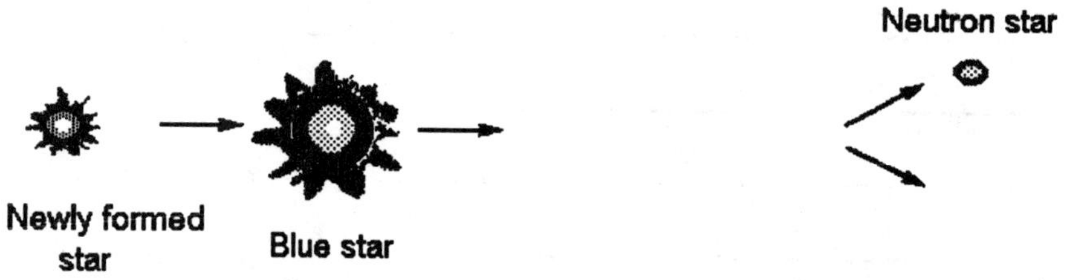

10. The diagram on the left shows the Sun, Earth, and Mars. Complete the diagram on the right by showing the approximate positions of Earth and Mars one year later.

Length of Year
Earth: 365 Earth days
Mars: 687 Earth days

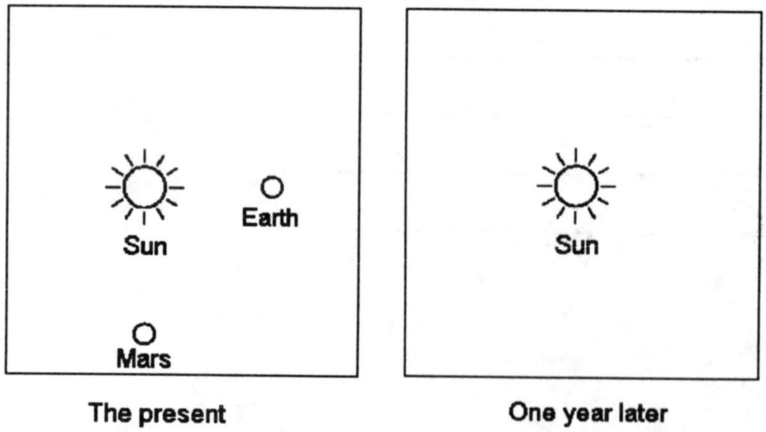

11. Students of Jamison Middle School were asked to make drawings of observations of celestial objects. Complete the drawing by showing what their next observation might be.

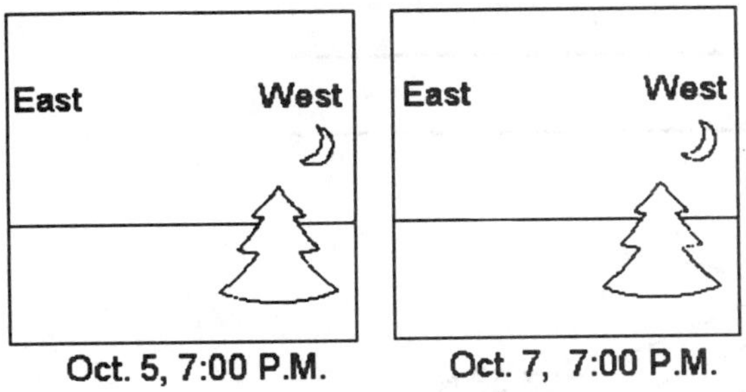

Illustrative

12. What is the difference between the heliocentric and geocentric models of the solar system? Draw two diagrams that show the difference between these models.

13. Consider the diagram below and answer these questions:

a. What shape is the solar system? Use the term *plane of the ecliptic* in your response.

____

____

____

b. Where does our solar system fit into the universe? In other words, what is our solar system's cosmic address?

____

____

____

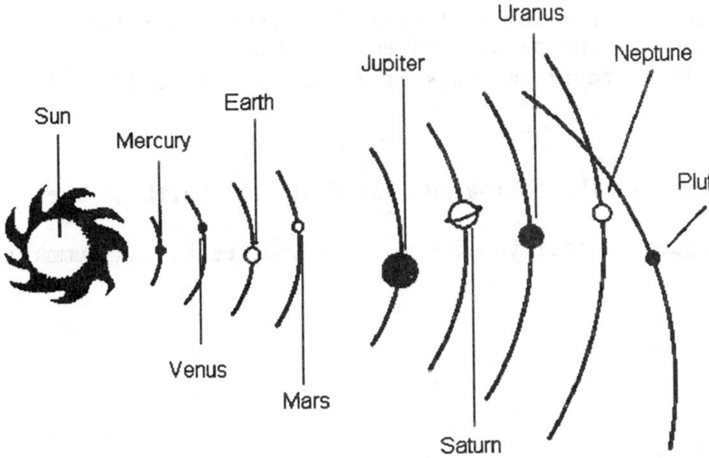

14. In each illustration, a stick is positioned vertically on the Earth at a point where it will not cast a shadow. What is the date for each illustration?

a. ______________________________

b. ______________________________

c. ______________________________

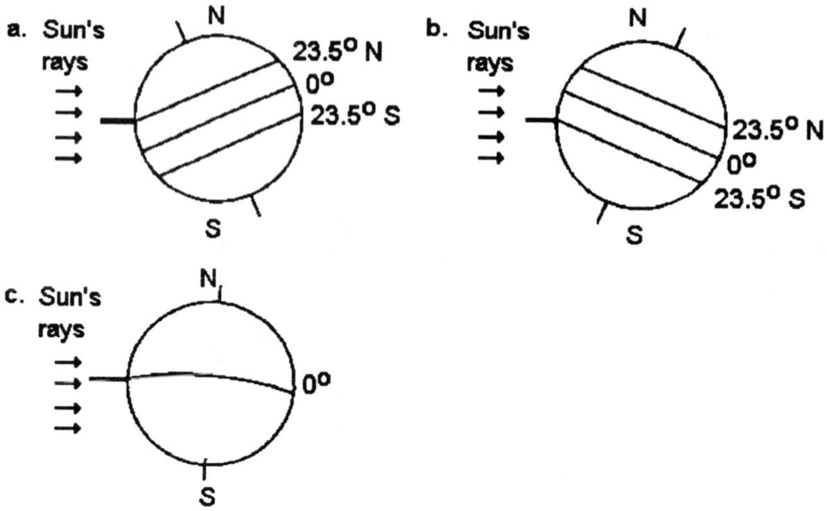

Performance Task

15. Collect data to prove or disprove the following statements.
    a. The Moon is in the same position in the sky at the same time each night.
    b. The Sun rises at a different point in the sky each day.

16. **Materials**
    - 3 plastic-foam balls
    - 2 toothpicks
    - a thumbtack

    Tasks
    a. Use one plastic-foam ball to represent the Sun and another to represent the Earth. Use two toothpicks stuck in opposite sides of the Earth to represent the Earth's axis. Now demonstrate why Earth has seasons. Show the relationship between the Sun and the Earth when it is summer in North America.
    b. Place a thumbtack in the Earth and show where it would be located in relation to the Sun in early evening, the middle of the night, and early morning.
    c. Use the third plastic-foam ball to represent Mars at a location where it would be visible in our night sky.

17. Prepare to debate either for or against the following statement: The Earth is the center of the universe.
    Gather all of the evidence and arguments that you can to make your case, and summarize your arguments in a one-page paper.

Answers to Unit 7 Extra Assessment Items

Word Usage

1. Answer:
   Sample answer: The fact that the universe appears to be *expanding* is used to support the *big-bang* theory of the *universe's* origins.

2. Answer:
   Sample answer: A *satellite* is any body that revolves about another so the *planets* are *satellites* of the Sun.

3. Answer:
   Sample answer: *Meteorites* reach the surface of the Earth and may cause *craters* upon impact. *Meteors*, on the other hand, burn up before hitting the ground.

Short Essay

4. Answer:
   Students' articles should include the following points:
   - A light-year is a measure of distance. It is the distance that light travels in 1 year.
   - Light from a source that is 3 billion light-years away takes 3 billion years to reach us. Therefore, the object is seen as it appeared 3 billion years ago.

5. Answer:
   Answers will vary. For example, for headline #1, *Amateur Astronomer Discovers New Comet*, the writer should inform the audience about the nature of comets. Points that could be mentioned include the following:
   - Comets are made of ice and dust.
   - Many comets have orbits that bring them close to the Sun.
   - The time of a comet's return can be predicted.
   - The tail of a comet is formed by dust and vapor blown back by the solar wind. The vapor glows because the dust reflects the sunlight.

6. Answer:
   There are many possible answers, including stars and galaxies. Consider the answer acceptable if one piece of information or a sketch is provided for each of two components of the universe.

7. Answer:
   a. Sample questions: (1) What is a black hole? (2) How did the universe form? (3) Why don't we always see the Moon in the sky? (4) Why are there rings around Saturn? (5) Could life as we know it exist on Mars?

   b. If these questions are answered by using information from this unit, consider the answer acceptable. Sample answers: (1) A black hole is a small, dense point in space that is formed from a massive, collapsing star. Its gravitational forces are so strong that nothing, not even light, can escape. (2) According to the big-bang theory, a cosmic event formed the universe as we know it 10 to 20 billion years ago.

Short Response

8. Answer:
The big-bang theory is one attempt to explain the origin of the universe. The theory states that the universe began 10 to 20 billion years ago with a cosmic event. According to this theory, the universe has not always existed as we see it today, but was once much smaller; thus, the universe has been expanding ever since it was formed.

Correction/Completion

9. Answer:

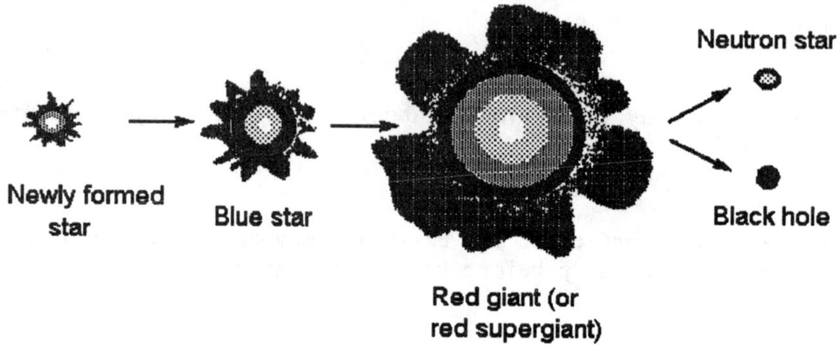

10. Answer:
The diagram should show the Earth in about the same position as in the first diagram. Mars should be about halfway around the Sun from its original position.

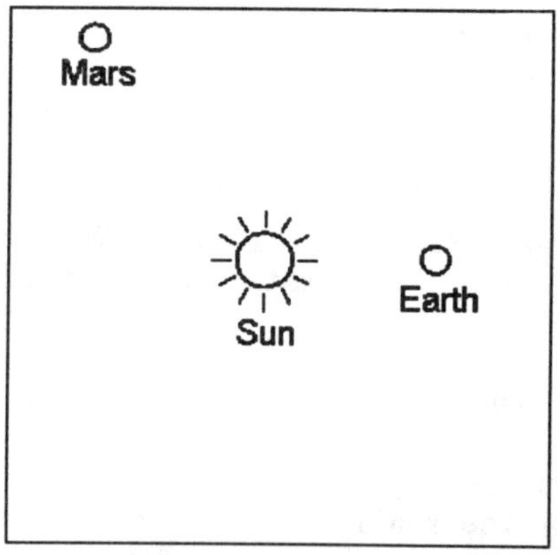

One year later

11. Answer: The diagram should show the Moon farther to the east, higher, and fuller.

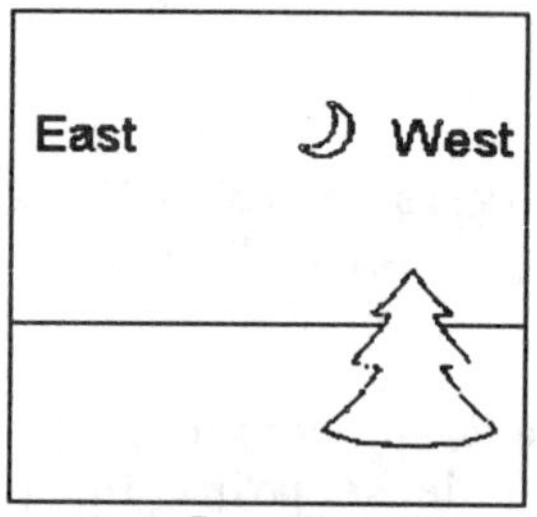

Oct. 9, 7:00 P.M.

Illustrative

12. Answer:

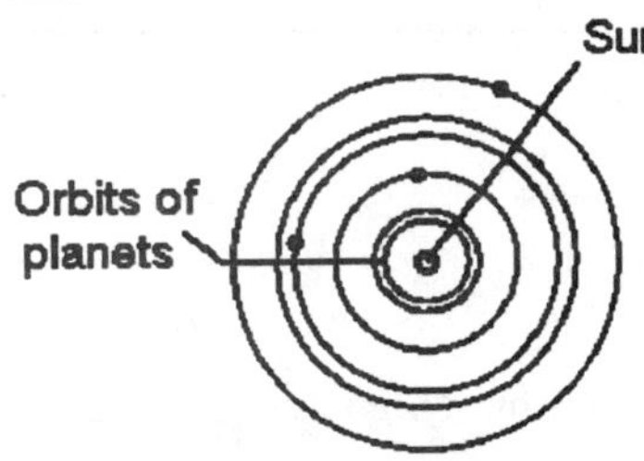

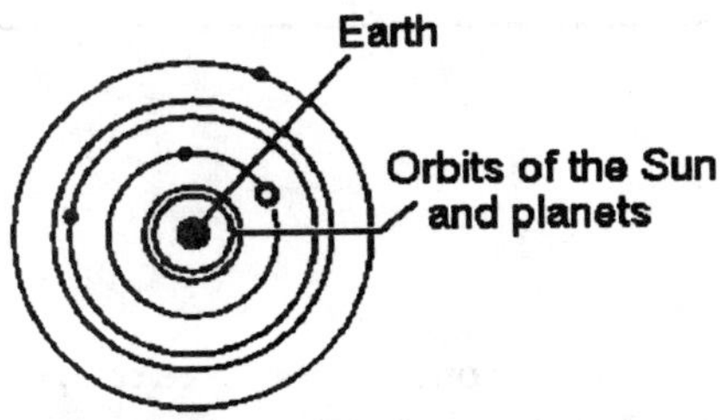

13. Answer:
   a. The solar system is flat because most of the planets lie in the same plane, called the *plane of the ecliptic*, as the Sun.
   b. Our solar system is a part of the Orion Arm of the Milky Way galaxy, which is part of a cluster of galaxies called the Local Group.

14. Answer:
   a. June 20 or 21, the summer solstice
   b. December 21 or 22, the winter solstice
   c. September 22 or 23, the autumnal equinox (Partial credit can be given for March 20 or 21, the vernal equinox.)

Performance Task

15. Answer:
   a. Data should be collected by observing the Moon in the sky at the same time each night and by carefully recording its position. Diagrams would be especially helpful.
   b. Data should be collected by observing the rising Sun (without looking directly at the Sun) for a fixed period of time. At sunrise each day, students should record the time when the Sun first appears above the horizon (or the top of a building). Be sure that students view the Sun from the same position each time. Students may wish to use some fixed object, such as a tree, to determine the exact point where the Sun rises. All observations should be carefully recorded. Diagrams are especially helpful.

16. Answer:
   Students should use the materials to effectively demonstrate their understanding of the orbits of Earth and Mars around the Sun and of the factors that cause seasonal variations on Earth.

17. Answer:
   Students should demonstrate the ability to develop a solid position, well-supported by evidence, for or against the statement. Their position should be clearly stated in a one-page paper and students should be able to defend their position in an oral debate.

Chapter 23 Assessment

Word Usage

1. Use all of the following terms in one or two sentences to show how they are related:
   a. soil, plants, water, nutrients, support

   ____

   ____

   ____

   b. soil, sand, silt, clay, loam

   ____

   ____

   ____

Correction/Completion

2. The statements below are either incorrect or incomplete. Your challenge is to make them correct and complete.
   a. Soil often contains living things called humus.

   ____

   ____

   ____

   b. Percolation rate tells how much water can be held by a soil material such as sand or silt.

   ____

   ____

   ____

Short Response

3. Suppose you were experimenting to see which type of soil was best for growing tomatoes. Describe at least three variables you should control in your experiment, and tell about three things you could measure or observe to determine how well the plants grew.

____

____

____

____

____

Illustrative

4. The illustration shows one possible layout for a biosphere.

   a. Label the illustration in as much detail as you can.
   b. Why does a biosphere need so many different types of environments?

____

____

____

____

____

Data for Interpretation

5. Hans conducted an experiment and recorded his data in the table below. Examine his data and answer the questions that follow.

Plant	Soil	Amount of water per day (mL)	Temp. (°C)	Amount of sunlight per day (hr.)	Height—day 1 (cm)	Height—day 15 (cm)
bean	loam	100	22	0	5	6
bean	loam	100	22	1	5	7
bean	loam	100	22	2	5	9
bean	loam	100	22	4	5	12
bean	loam	100	22	8	5	16

a. What do you think was the purpose of this experiment?

____

____

____

b. What can he conclude from his data?

____

____

____

Answers to Chapter 23 Assessment

Word Usage

1. Answer:
   a. Sample answer: *Soil* enables *plants* to grow by providing *support* and by holding *water* and other *nutrients* that the plants absorb through their roots.
   b. Sample answer: *Loam*, a type of *soil*, is a mixture of *sand*, *silt*, and *clay*.

Correction/Completion

2. Answer:
   a. Soil often contains *the decayed remains of* living things, called humus.
   b. Percolation rate tells how *quickly water passes through* a soil material such as sand or silt.

Short Response

3. Answer:
   Possible controls include using the same seeds, giving all plants equal water and sunshine, and leaving the same amount of space between all plants. Possible observations include the height and mass of the plants, the number and size of the leaves, and the mass and taste of the tomatoes.

Illustrative

4. Answer:
   a. See labeled illustration below.
   b. Sample answer: Organisms, including plants and animals, will be most successful when a certain level of diversity is maintained. That means that various resources are needed to support different kinds of life. A biosphere with different types of environments will be more likely to provide the living conditions needed by various living things.

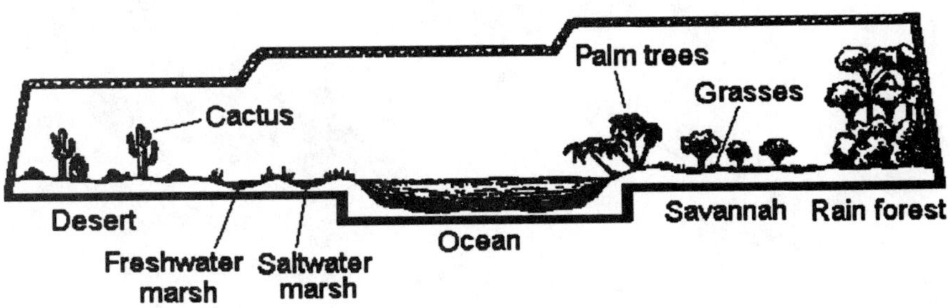

Difficulty: 3

Data for Interpretation

5. Answer:
   a. Hans wanted to determine the effect of sunlight on the growth of bean plants in loam.
   b. Hans can conclude that bean plants planted in loam, kept at 22°C, and given 100 mL of water each day grow best when given 8 hours of sunlight each day.

   Difficulty: 4

Chapter 24 Assessment

Word Usage

1. a. The prefix *geo* means having to do with the Earth, and the word *geoponic* means having to do with agriculture. Based on this definition, how would you define the word *hydroponics*?

   ____

   ____

   ____

   ____

   ____

   b. Define the term *osmosis* using the following words: concentrated, water, more, less.

   ____

   ____

   ____

   ____

   ____

Correction/Completion

2. The following statements are either incorrect or incomplete. Your challenge is to make them correct and complete.
   a. Transpiration is the movement of water into a plant through its roots.

   ____

   ____

   ____

   b. During pollination, pollen from the tip of a perfect flower's pistil is transferred to the tip of the stamen of a different plant.

   ____

   ____

   ____

Short Response

3. Match each method of vegetative reproduction with the example of that method.

a. division	___	Ivy spreads when its stem touches the ground and grows new roots.
b. layering	___	Mrs. Wiley snipped off a few pieces of her coleus plant and repotted them to grow new plants.
c. cutting	___	A farmer transferred twigs from his best peach tree to the branches of his less-productive peach trees.
d. grafting	___	Faye grows new irises by digging up, separating, and replanting the roots of existing irises.

Numerical Problem

4. A farmer plants 30,000 spinach seeds with a germination rate of 65 percent. If 75 percent of the germinated seeds develop into healthy mature plants, and each mature plant transpires 75 L of water before harvest time, how much water will the mature plants transpire? Show your work.

____

____

____

Short Essay

5. Suppose that you are trying to rejuvenate an old apple orchard that had been neglected for years. After carefully pruning all of the dead branches, you find that each tree has a few healthy branches left but that the apple crop is very small. Compare the method of grafting with planting new seeds or saplings to increase the apple crop.

____

____

____

____

____

____

____

____

Answers to Chapter 24 Assessment

Word Usage

1. Answer:
   a. Sample answer: Geoponics is a method of growing plants in soil on the Earth. *Hydro* means having to do with water, so *hydroponics* is a method of using solutions of water and certain nutrients to grow plants.
   b. Sample answer: *Water* moves through cell membranes from areas where water is *more concentrated* to areas where water is *less* concentrated.

Correction/Completion

2. Answer:
   a. Transpiration is the movement of water *out of* a plant through its *leaves*.
   b. During pollination, pollen from the tip of a perfect flower's *stamen* is transferred to the tip of the *pistil* of *the same* plant.

Short Response

3. Answer:
   b Ivy spreads when its stem touches the ground and grows new roots.
   c Mrs. Wiley snipped off a few pieces of her coleus plant and repotted them to grow new plants.
   d A farmer transferred twigs from his best peach tree to the branches of his less-productive peach trees.
   a Faye grows new irises by digging up, separating, and replanting the roots of existing irises.

Numerical Problem

4. Answer:
   30,000 x .65 = 19,500 germinated seeds
   19,500 x .75 = 14,625 mature spinach plants
   14,625 x 75 L = 1,096,875 L of water

   Difficulty: 3

Short Essay

5. Answer:
   Sample answer: If you start new trees from seeds or from saplings, it will take many years for the trees to grow and even longer for them to bear fruit. However, the trees in the old orchard have established trunks and root systems. By successfully grafting new branches onto the existing limbs, you will soon have productive trees again.

   Difficulty: 4

Chapter 25 Assessment

Word Usage

1. For each group of words below, write one or two sentences to show how they are related.
   a. slugs, aphids, pesticides, alternatives, organic gardeners

   b. landscape, plants, light, drainage, design

Correction/Completion

2. The following statements are either incorrect or incomplete. Your challenge is to make them correct and complete.
   a. A low-lying area on a piece of property is a good place for a vegetable garden.

   b. Ladybugs are pests and should be killed with chemical pesticides.

Short Response

3. Micah has a 5 m x 5 m garden plot in which he grows lettuce, beans, and squash. He is trying to decide whether to use the two unplanted rows to grow corn or tomatoes. He likes to eat both. What advice would you give Micah? Explain your answer.

____

____

____

____

Graphic

4. This graph shows results of a study on the rate of photosynthesis in a leaf.
   a. What two factors are being compared?

   ____

   b. What happens to the rate of photosynthesis in the leaf as the amount of light increases?

   ____

   c. How could a graph like this be useful to a landscape designer?

   ____

   ____

   ____

   ____

   ____

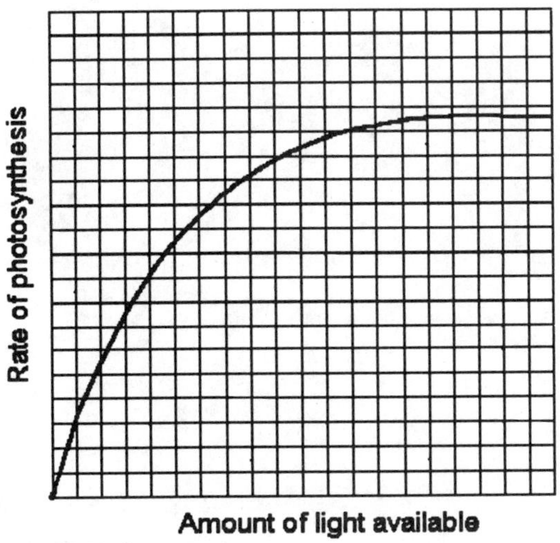

Short Essay

5. "Companion planting" is a technique used by some organic gardeners. One example is to alternate a row of marigolds between each row of tomato plants in a garden. What do you think is the purpose of companion planting, and why do you think it works?

Answers to Chapter 25 Assessment

Word Usage

1. Answer:
   a. Sample answer: Instead of using *pesticides* to kill *slugs*, *aphids*, and other organisms that are harmful to plants, *organic gardeners* look for safer *alternatives*.
   b. Sample answer: When making a *landscape design*, one should consider features such as amount of *light* and *drainage* to help decide where different types of *plants* belong.

Correction/Completion

2. Answer:
   a. Sample answer: A low-lying area on a piece of property is a good place for a *pool or a pond*.
   b. Sample answer: Ladybugs *eat* pests and should *not* be killed *because they can replace* chemical pesticides.

Short Response

3. Answer:
   Sample answer: It might be better for Micah to plant tomatoes because they take up less space and yield more produce per plant. Also, tomatoes are generally more expensive than corn in stores, so he would save money.

Graphic

4. Answer:
   a. The rate of photosynthesis and the amount of light available
   b. It increases directly with the amount of light available.
   c. Sample answer: This graph would give insight into the amount of light needed by this particular plant to successfully produce food for itself. The landscape designer could use this information to determine where the plant should be located.

   Difficulty: 3

Short Essay

5. Answer:
   Sample answer: The purpose of companion planting is to protect plants from insects and other harmful organisms without using pesticides and other chemicals that can harm people or pets. Companion planting works in this case, for example, because some of the pests that generally attack tomatoes avoid marigolds, so by alternating the rows, a gardener can protect his or her tomato plants.

   Difficulty: 4

Unit 8 End-of-Unit Assessment

Word Usage

1. a. Using the words *aggregate*, *hydroponics*, *nutrient solution*, and *support*, write brief directions for growing plants without soil.

   ____

   ____

   ____

   ____

   ____

   b. Use the words *humus*, *loam*, and *nutrients* in one or more sentences that describe good garden soil.

   ____

   ____

   ____

   ____

2. a. Chen and Sue bought an older home that had been vacant for awhile. In the backyard, they found many weeds, various animal pests, and some signs of mold on flowering plants. Use the following words as you explain how they improved the conditions in their backyard: pesticides, research, alternatives.

   ____

   ____

   ____

   ____

   b. Steve and Jan are helping a group of handicapped children who enjoy growing plants but cannot afford to buy a lot of new plants from a nursery. Write one or more sentences to show what they can do, using the following words: seeds, cuttings, vegetative reproduction.

   ____

   ____

   ____

   ____

Correction/Completion

3. For each blank, fill in the word that makes the sentence correct.

   a. After fertilization, the ovary of a peach blossom grows to become the ________, and the egg cells develop into the ________.

   b. A seed contains the ________ of a new plant.

   c. Water enters through the ________ of a plant and moves from cell to cell by a process called ________. Water moves from a place where it has a ________ concentration to a place where it has a ________ concentration.

4. The following statements were made by students studying percolation rate. Beneath each comment, give a response to correct the student's misconception.

   a. Annie: "Percolation rate is the amount of water that stays in the soil, isn't it?"

   ____

   ____

   ____

   ____

   b. Jon: "Then why is it the sandy soil had the highest percolation rate? It didn't hold much water. But the clay sample stayed wet."

   ____

   ____

   ____

   ____

   c. Corey: "Clay must be the best for gardens, then."

   ____

   ____

   ____

   ____

Short Response

5. Your plants are looking pale and yellowish, but you know they have enough water and sunlight. A friend suggests that you need a fertilizer with a special N-P-K rating.

   a. What does N-P-K refer to?

   b. How can you use this information to help your plants?

6. Identify the role that each of these terms has in plant reproduction.

   a. pollination

   b. vegetative reproduction

Illustrative

7. Use the two pictures to answer the questions.
   a. Draw arrows inside the illustrations to indicate the direction that the water moves.
   b. Identify the process occurring in the beakers. How does this process work in plants?

___

___

___

___

___

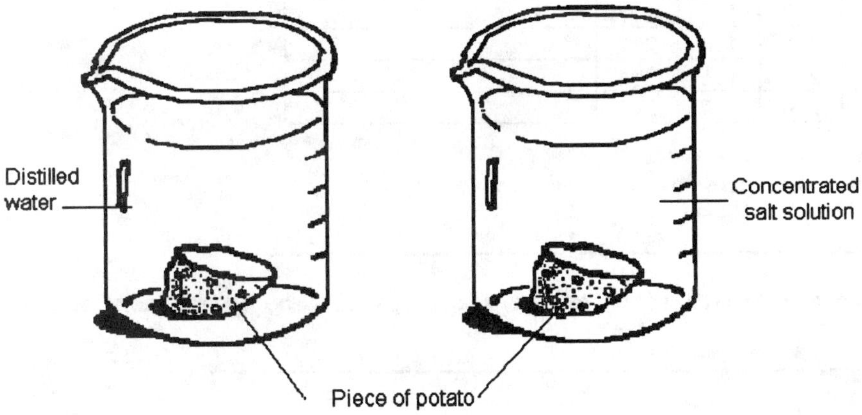

Data for Interpretation

8. Fran and Mike planted five radish seeds. When the seedlings first appeared above the soil (day 1), they put the containers on the windowsill. Then they began a daily log of observations about growth. A table from their ScienceLog is given below. Complete the table.

Day	Plant height (cm) A	B	C	D	E	Average height (cm)
1	0.4	0.5	0.4	0.3	0.5	0.42
2	0.5	0.6	0.6	0.4	0.7	0.56
3	0.7	0.8	0.8	0.6	0.9	0.76
4	0.8	1.0	1.0	0.7	1.2	0.94
5	1.0	1.2	1.3	0.9	1.3	1.14
6	1.3	1.4	1.5	1.1	1.5	
7	1.6	1.7	1.8	1.3	1.7	
8	1.8	2.0	2.1	1.4	2.1	
9	2.1	2.3	2.4	1.5	2.4	
10	2.4	2.5	2.6	1.7	2.6	

a. During which days was the growth rate the greatest?

____

b. Predict what the average growth rate for the seedlings will be in the next few days.

____

____

____

____

Short Response

9. Why do trees need leaves?

____

____

____

____

____

____

Short Essay

10. Give step-by-step instructions that explain how to start a new plant from a leaf cutting. Name the method of reproduction by supplying a title for the directions.

____

____

____

____

____

____

____

____

____

Illustrative

11. **Chelsea:** I'm not sure about the parts of flowers and what they're for.

Nick: Do you mean, like the stamen and the anthers?

Bradley: Those things belong together, don't they? Where's the style? Does that hold the anthers up?

Camilla: No—you have it wrong. The style is part of the pistil and the filament holds the anther. Or is it the stamen?

Chelsea: Then where does the pollen come from?

Help these students understand. Draw and label the parts of a flower. Include petals and sepals in your sketch.

Short Essay

12. Suppose that your garden store sells three kinds of fertilizers. You want to find out which kind to use to grow the greatest amount of beans. Describe the procedure you would follow to determine which kind of fertilizer is best for your purpose.

Answers to Unit 8 End-of-Unit Assessment

Word Usage

1. Answer:
   a. Sample answer: Plants grown by *hydroponics* are placed in an *aggregate* of small stones or sand for *support* and are fed a *nutrient solution* that contains all of the minerals that a plant needs to survive.
   b. Sample answer: *Loam* is an excellent soil for a garden because it is rich in *humus*, which contains many *nutrients* that help plants grow.

2. Answer:
   a. Sample answer: Instead of using *pesticides*, which might harm people or animals, Chen and Sue did some *research* and decided to use safer *alternatives* to control the pests, weeds, and mold.
   b. Sample answer: Steve and Jan could help the children grow new plants from the *seeds* of other plants, or they could use *cuttings* that will become whole plants through *vegetative reproduction*.

Correction/Completion

3. Answer:
   a. After fertilization, the ovary of a peach blossom grows to become the _fruit_, and the egg cells develop into the _seeds_.
   b. A seed contains the _embryo_ of a new plant.
   c. Water enters through the _roots_ of a plant and moves from cell to cell by a process called _osmosis_. Water moves from a place where it has a _higher_ concentration to a place where it has a _lower_ concentration.

4. Answer:
   a. Percolation rate is the time it takes for a known volume of water to pass through a soil sample.
   b. Sandy soil lets water pass through quickly, which results in a high (fast) percolation rate. A clay sample holds much more water and lets water pass through very slowly.
   c. Plants need air spaces in soil, and wet clay cannot supply the needed air. The best soil to use is loam, which consists of a mixture of soil particles, humus, air, and water.

Short Response

5. Answer:
   a. N-P-K refers to the three chemical nutrients that plants need in the greatest quantities. N stands for nitrogen, P stands for phosphorus, and K stands for potassium.
   b. Sample answer: By knowing that the numbers for N-P-K indicate the percentage of each chemical in a fertilizer, you can choose a fertilizer that will supply the right amount of nutrients for your plants.

6. Answer:
   a. Sample answer: Pollination occurs when pollen from the anthers of a flower are carried to the pistil of the same or a different flower. The pollen moves down the pistil to reach the ovary and fertilize the egg.
   b. Sample answer: Under controlled conditions, a part of an adult plant, such as a leaf, root, or stem, can be removed and replanted so that a new plant will grow. This is vegetative reproduction.

Illustrative

7. Answer:
   a. See illustration below.
   b. The process is osmosis—the movement of water from areas of high concentration to areas of low concentration. In plants, osmosis occurs when plants absorb water and nutrients from the soil, move water through their stems and leaves, and release water as vapor after photosynthesis.

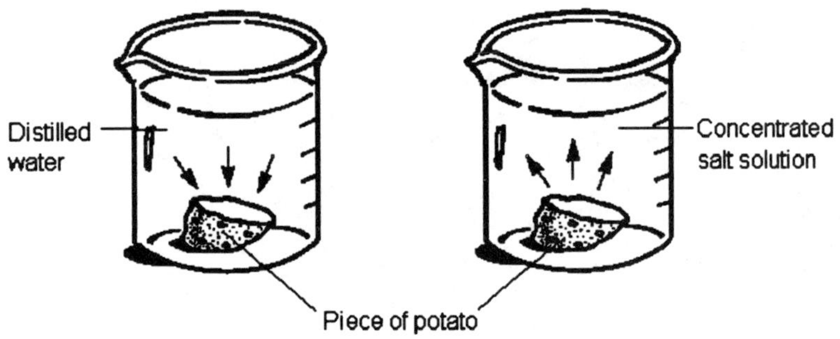

Data for Interpretation

8. Answer:

Day	Plant Height (cm) A	B	C	D	E	Average height (cm)
1	0.4	0.5	0.4	0.3	0.5	0.42
2	0.5	0.6	0.6	0.4	0.7	0.56
3	0.7	0.8	0.8	0.6	0.9	0.76
4	0.8	1.0	1.0	0.7	1.2	0.94
5	1.0	1.2	1.3	0.9	1.3	1.14
6	1.3	1.4	1.5	1.1	1.5	*1.36*
7	1.6	1.7	1.8	1.3	1.7	*1.62*
8	1.8	2.0	2.1	1.4	2.1	*1.88*
9	2.1	2.3	2.4	1.5	2.4	*2.14*
10	2.4	2.5	2.6	1.7	2.6	*2.36*

a. Growth was the greatest during days 6 through 10.
b. Sample answer: The growth rate will probably slow down and stabilize over the next few days.

Short Response

9. Answer:
Sample answer: Trees need leaves for photosynthesis and transpiration. In photosynthesis, food is made from carbon dioxide, water, and energy from the sun. Through transpiration, water moves out of a plant by evaporating from its leaves. This helps water move through the plant.

Difficulty: 3

Short Essay

10. Answer:
Sample answer: Vegetative Reproduction
Remove a leaf with 2 to 3 cm of stem attached. Put the stem of the leaf in moist soil or potting mixture. Cover the plant pot and the leaf with a clear plastic bag until small leaves begin to form where the stem enters the soil. By this time strong roots should be developing. Remove the plastic bag. Later, cut away the old leaf and put the new plant into a pot of general-purpose soil mixture.

Difficulty: 3

Illustrative

11. Answer: Sample sketch:

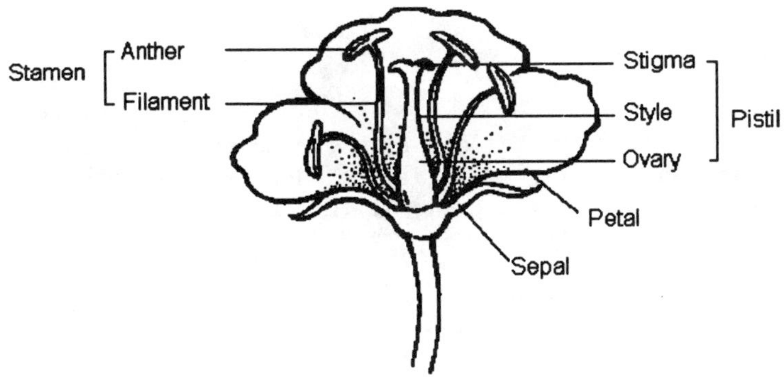

Difficulty: 4

Short Essay

12. Answer:
In comparing the effects of the three kinds of fertilizers, all of the other variables must be kept the same. This means using the same amount of fertilizer, the same kind and amount of soil, and the same amount of water and light for each plant or group of plants. The answer is acceptable if it mentions at least two variables that must be kept the same. If it mentions that all of the variables are to be kept the same except the kind of fertilizer, that the experiment should be repeated to check the results, or that each fertilizer should be applied to a number of bean plants, give the answer a bonus.

Difficulty: 4

Unit 8 Activity Assessment

Activity Assessment

1. **Gardens of Food**

Teacher's Notes

Overview
Students identify and analyze plant parts used as food. They create a classification system for the plant parts.

Materials
(per activity station)
- 2 pieces of fruit (such as apple, tomato, orange, lemon, pear, etc.)
- 2 edible roots (such as carrot, sweet potato, beet, etc.)
- 2 edible leaves (such as lettuce leaf, cabbage leaf, bay leaf, parsley, etc.)
- 2 edible stems (such as asparagus, broccoli, etc.)
- 2 samples of edible seeds (such as sunflower, peanut, pecan, walnut, pumpkin, etc.)
- 1 knife
- paper towels
- poster board and colored pencils

Preparation
Prior to the assessment, equip student activity stations with the materials needed for each experiment. Mix up the items so that similar items are not next to each other. Remind students that they should not eat any of the items at the activity station.

Time Required
Each student should have 25 minutes at the activity station and 15 minutes to complete the Data Chart.

Performance
At the end of the assessment, students should turn in the following:
- a completed Data Chart
- a classification system on poster board

Evaluation
The following is a recommended breakdown for evaluation of this Activity Assessment:
- 30% accurate identification of the main characteristics of each plant part
- 35% accurate identification of the plant part and its function
- 35% classification of the plant parts

Safety Alert!

2. Gardens of Food

As you probably know, fruit and vegetables are an important part of a balanced diet. At your activity station, you'll find some items that are commonly eaten. Each of these items is part of a plant. Keeping that in mind, examine each item carefully, and note its various characteristics. Then decide what function each food source provides for the plant it came from. Finally, group the items according to similar characteristics.

Before You Begin . . .
As you work through the tasks, keep in mind that your teacher will be observing:
- how you identify the main characteristics of each plant part
- your analysis of the plant parts and their functions
- how clearly and logically you classify the plant parts

Dig in!

Task 1 Examine each item carefully, noting its characteristics in your Data Chart. Consider color, shape, texture, internal structure, and any other important features of each item. Do not taste any of the items.

Task 2 Identify which part of the original plant the sample came from. Then decide what function it has in a complete plant. Record your answers in the Data Chart.

Task 3 Using your Data Chart as a guide, classify the plant parts. Record your classification system on the poster board.

3. **Data Chart**

Name of sample	Characteristics	Part of original plant	Function in plant

Answers to Unit 8 Activity Assessment

1. Answer: Not applicable (teacher's notes)

2. Answer: Not applicable (student's notes)

3. Answer:
   Answers will vary, but student observations and classification systems should be clear, concise, and well-presented.

Unit 8 SourceBook Assessment

Multiple Choice

1. _______ are the food-making organs of plants.
   a. Leaves b. Stems c. Roots d. Flowers

2. The _______ are made of tissues that bring water into the leaf.
   a. veins b. blades c. petioles d. flowers

3. The organelle where photosynthesis takes place is called the
   a. stoma. b. nucleus. c. chloroplast. d. None of the above

4. Stomata allow carbon dioxide and water vapor to go into and out of the leaf, and they are located on the (top/bottom) of the leaf.
   (Circle the correct word.)

5. What causes the guard cells to open a leaf's stomata?
   a. water b. light c. carbon dioxide d. oxygen

6. Which of the following are stems?
   a. potato b. tree trunk c. asparagus stalk
   d. All of the above e. b and c only

7. Which of the following is not true about roots?
   a. Roots are much smaller than the part of the plant above the ground.
   b. Roots hold the plants in the ground.
   c. Roots collect water and minerals for the plant.
   d. All of the above are true.

8. The type of root system that would survive best in a climate that has short, hard rains would be a
   a. fibrous root system. b. taproot system.

9. Roots grow in length by making new cells where?
   a. at the beginning b. in the middle c. at the end

10. Water is taken into root hairs of a plant by
    a. absorption. b. osmosis. c. respiration. d. dilution.

11. Which of the following is not true about cellulose?
    a. It can be digested by humans.
    b. It is the material that makes up a plant's cell walls.
    c. Cotton contains it.
    d. It is used to make $10 bills.

12. The petiole of a plant is part of its
    a. leaf. b. stem. c. root. d. flower.

13. Edible plant parts include
    a. roots. b. stems. c. leaves. d. flowers.
    e. fruits. f. seeds. g. a and d h. All of these

14. Which of the following is not a fruit?
a. wheat b. beet c. tomato d. banana

Short Response

15. Fill in the blanks of the photosynthesis reaction with the correct letters.
a. water b. sugar c. oxygen d. carbon dioxide

________ + ________ → ________ + ________

16. In the vascular bundles of a plant's stem, the large, thick-walled cells carry ________, and the smaller, thin-walled cells carry ________. (Fill in each blank with the correct letter.)
a. water b. food

True/False

17. The leaves of a plant are covered with an outer layer called the epidermis.
a. true b. false

18. Water exits a plant through exfoliation.
a. true b. false

19. The bark of a tree is made up of a layer of dead stem cells.
a. true b. false

20. When a stigma receives pollen from the anther of the same flower, cross-pollination takes place.
a. true b. false

Multiple Choice

21. Aspirin contains salicylic acid, which originally came from what tree?
a. yew b. willow c. oak d. cottonwood

Short Response

22. If the soil around a plant becomes very dry, what happens to the process of photosynthesis?

____

____

____

23. Why might many AIDS researchers be concerned about the destruction of tropical rain forests?

____

____

____

____

____

____

____

____

____

24. Why is it important not to pull leaves off of shrubs?

____

____

____

____

Matching

25. Match each food item on the right with the plant part on the left that it represents.

a. leaf ________ celery

b. stem ________ lettuce

c. root ________ peanut

d. seed ________ turnip

Answers to Unit 8 SourceBook Assessment

Multiple Choice

1. Answer: a. Leaves

2. Answer: a. veins

3. Answer: c. chloroplast.

4. Answer: bottom

5. Answer: a. water

6. Answer: d. All of the above

7. Answer: a. Roots are much smaller than the part of the plant above the ground.

8. Answer: a. fibrous root system.

9. Answer: c. at the end

10. Answer: b. osmosis.

11. Answer: a. It can be digested by humans.

12. Answer: a. leaf.

13. Answer: h. All of these

14. Answer: b. beet

Short Response

15. Answer: __a__ + __d__ → __b__ + __c__

16. Answer:
In the vascular bundles of a plant's stem, the large, thick-walled cells carry __a__, and the smaller, thin-walled cells carry __b__.

True/False

17. Answer: a. true

18. Answer: b. false

19. Answer: a. true

20. Answer: b. false

Multiple Choice

21. Answer: b. willow

Short Response

22. Answer: It slows down or stops.

23. Answer:
Answers may vary but should include this main idea: Many medicines come from plants and trees, and many plants are becoming extinct before they have been researched for their medicinal value. An AIDS researcher might fear that with the rapid destruction of the abundant plant resources in tropical rain forests, a plant that could help cure AIDS might disappear before it could be discovered.

24. Answer:
Answers may vary but should include this main idea: It is in the leaves that most of the plant's photosynthesis takes place. With fewer leaves, less photosynthesis occurs, and the plants will not grow as well.

Matching

25. Answer:
<u>b</u> celery
<u>a</u> lettuce
<u>d</u> peanut
<u>c</u> turnip

Unit 8 Extra Assessment Items

Word Usage

1. Imagine that you had to start a garden on Mars. Use the following words to explain what you must consider as you begin growing your garden: biosphere, loam, nutrients, percolation rate, hydroponics, and rooting medium.

____

____

____

____

____

____

Correction/Completion

2. Insects are needed to pollinate all plants.
This statement is either incorrect or incomplete. Make the statement correct and complete.

____

____

____

____

3. One breeding technique involves the transfer of pollen from the stamens of one plant to the pistil of another to yield some desired features of both plants. This method is called ________________.

4. A male plant's flowers have no ______________________________,

while a female plant's flowers have no ______________________________.

5. **Clio**: I want to fertilize my African violets. The directions say to add 15 g of dry fertilizer to 1 L of water. Since my plants have not had fertilizer in a long time, I'm going to add 25 g or 30 g of fertilizer instead. That should help.
Mark: I think such a concentrated solution could be harmful because . . . .

Complete Mark's statement. Explain to Clio what could happen to her plants if she uses a fertilizer solution that's too strong. Why does this result occur?

___

___

___

___

Short Essay

6. Doug said, "I'll just use this pesticide called Kill 'Em All. That's the only way to get rid of garden pests!" Describe to Doug some possible dangers of using pesticides.

___

___

___

___

___

___

7. A friend of yours is interested in growing indoor plants but has very little experience. You know that one way to start a new plant is to use seeds. Explain what is needed for the germination of a seed to occur.

___

___

___

___

___

___

8. With words or an illustration or both, describe how to pollinate flowers indoors.

Short Response

9. List the factors that are necessary for plants to survive.

10. Cathy observed that raising the temperature in the room affected the growth of geraniums. What factors should be controlled in an experiment to confirm this observation? What would be the variable?

11. Audrey stored a group of seeds in a damp place and found that they had a lower germination rate than did seeds she stored in a dry place. What factors should be controlled in an experiment to confirm this observation? What would be the variable?

12. Pat found that the pH of the soil became more acidic when more compost was added. What factor should be controlled in an experiment to confirm this observation? What would be the variable?

13. List five things that animals and plants need to survive in a spacecraft.

____

____

____

____

14. Nick checked the N-P-K ratings of fertilizers in the garden shop. He needed a good general fertilizer. Mrs. Li suggested 20-20-20.
    a. What do N, P, and K refer to?

    ____

    ____

    ____

    b. What do the numbers refer to?

    ____

    ____

    ____

    c. What percentage of other substances is found in a fertilizer with an N-P-K rating of 20-20-20?

    ____

    ____

    ____

Graphic

15. This graph shows results of a study on the rate of food production in a leaf.

a. What happens to the rate of food production in the leaf as the light energy increases?

____

b. What happens to the rate of food production in the leaf as the percentage of carbon dioxide in the air increases?

____

c. Name one variable that has not been changed.

____

d. Predict how changing the variable in part (c) might have changed the results.

____

____

____

____

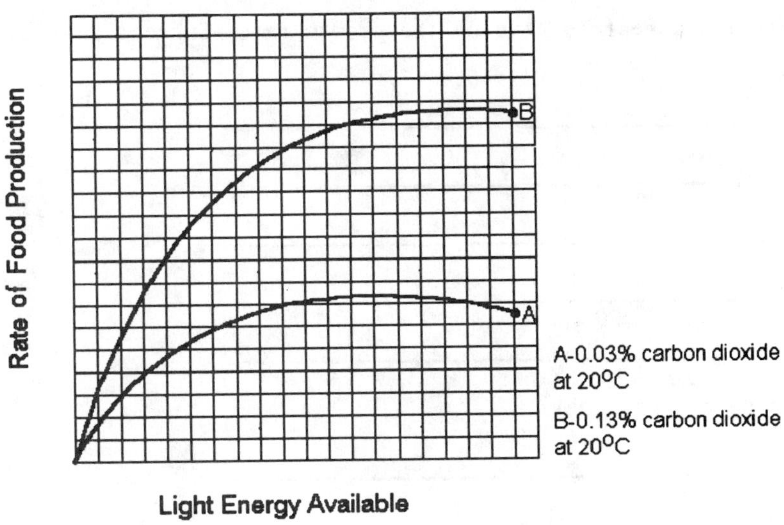

16. The following graph shows the results of experiments that tested the effect of air flow on the growth rate of plants. Using the graph, answer these questions.

a. With what rate of air flow did the plants have the fastest growth rate?

____

b. Did the growth rate increase, decrease, or remain the same when air flow was increased from 30 m to 60 m per minute?

____

c. What would you predict the growth rate would be with an air flow of 180 m per minute?

____

d. What effect did increasing air flow from 0 m to 30 m per minute have on the growth rate of plants? Suggest reasons for this effect.

____

____

____

____

____

e. What effect did faster air flow appear to have on the growth rate? Suggest reasons for this effect.

____

____

____

____

____

f. How could this information help a gardener?

____

____

____

____

____

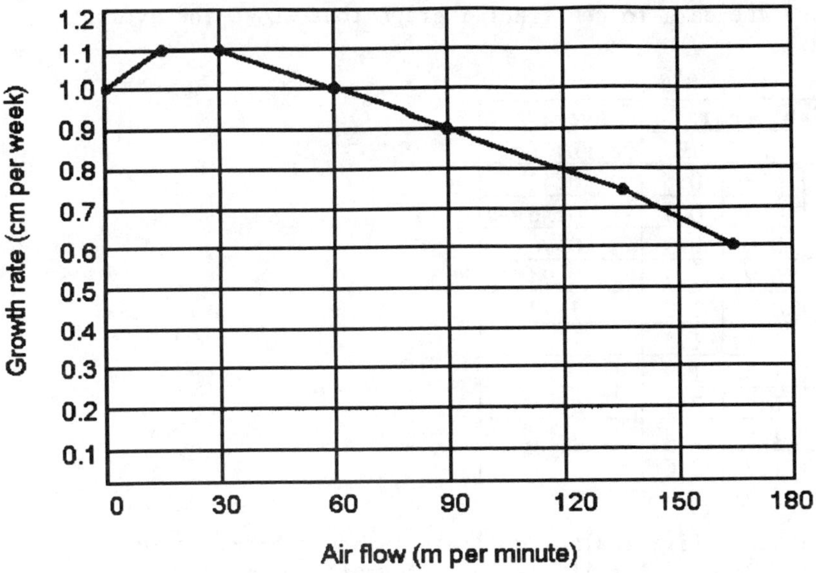
Growth rate (cm per week)
1.2
1.1
1.0
0.9
0.8
0.7
0.6
0.5
0.4
0.3
0.2
0.1
0
30
60
90
120
150
180
Air flow (m per minute)

17. a. Complete the table, and use the data to construct a graph that shows the average growth of radish seedlings.

Day	Plant height (cm)					Average height (cm)
	A	B	C	D	E	
1	0.4	0.5	0.4	0.3	0.5	0.42
2	0.5	0.6	0.6	0.4	0.7	0.56
3	0.7	0.8	0.8	0.6	0.9	0.76
4	0.8	1.0	1.0	0.7	1.2	0.94
5	1.0	1.2	1.3	0.9	1.3	1.14
6	1.3	1.4	1.5	1.1	1.5	
7	1.6	1.7	1.8	1.3	1.7	
8	1.8	2.0	2.1	1.4	2.1	
9	2.1	2.3	2.4	1.5	2.4	
10	2.4	2.5	2.6	1.7	2.6	

b. A table of the average height of five radish seedlings is shown here. These seedlings were treated with a liquid fertilizer. On the graph you made in (a), construct a line that shows the growth of these seedlings.

Day	1	2	3	4	5	6	7	8	9	10
Average height (cm)	0.5	0.5	0.9	1.3	1.8	2.3	2.8	3.4	3.9	4.4

c. What can you conclude about the effect of fertilizer on the growth of seedlings?

____

____

____

____

____

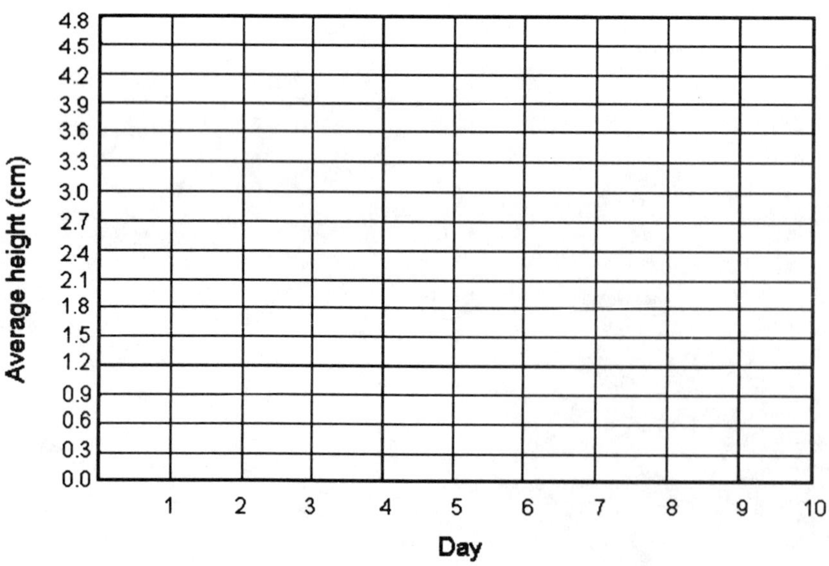

8. Using the grid that follows, construct a graph from the information in the table, which shows soil temperature at three depths over the course of 14 months. Use a different color to graph the data for each depth.

	Soil temperature (°C) at		
Month	3 cm	15 cm	75 cm
August	23	20	18
September	17	17	17
October	9	14	15
November	3	8	10
December	-1	4	7
January	-4	2	4
February	-4	0	3
March	-1	-1	2
April	3	0	2
May	12	8	5
June	20	15	12
July	22	18	16
August	21	19	17
September	17	16	16

Use the data in the table and your graph to answer the following questions:

a. During which 3 months are the temperatures of the soil most equal at all three depths?

____

b. During which 3 months do the temperatures cover the largest range?

____

c. Which depth has the greatest variation in temperature (from highest to lowest) throughout the year? Give a reason for this.

____

____

d. Which depth has the least variation in temperature throughout the year? Give a reason for this.

____

____

e. How would you expect the soil temperature to vary over the same 14-month time period at a depth of 150 cm (that is, how would the graph for this depth compare with the others)?

____

____

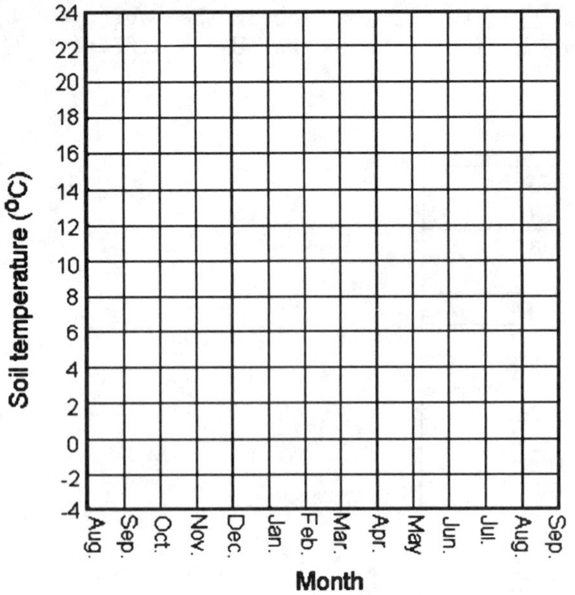

Illustrative

19. Using the words *pistil*, *stamen*, *petal*, *ovary*, and *pollen*, complete the diagram by labeling the parts of the flower. Indicate which parts are involved in pollination by circling their labels.

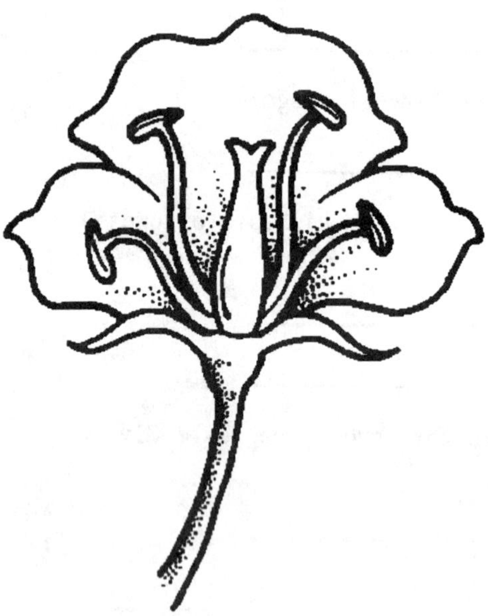

20. Explain how to grow a new plant using vegetative reproduction. Draw and label one or more sketches to illustrate your explanation.

21. Describe how a plant can be grown without soil. Provide a labeled diagram to accompany your description.

Performance Task

22. You are the marketing director for George's Garden-Gro products. Your challenge is to advertise the benefits of a new product, an organic fertilizer. The client would like you to use one of the following methods for advertising their new product: a TV jingle, a rap song, a radio or TV commercial, a flyer, a poster, an article for a seed catalogue or a gardening magazine. Research these options, and then present your ideas to the client (the class).

23. Find an indoor space that could be filled by some growing plants, and then plan how plants could be added to that space. Provide a detailed sketch of the area, showing where you would place at least 10 kinds of plants. Do research to find out how much space the plants need, how much light they need, etc. Then provide a written explanation of where you plan to place each plant.

24. A new homeowner has asked you to landscape the barren grounds around his home. Create a sketch of the area and indicate how plants could be added to the space. Include at least 10 kinds of plants in your plan. Do research to find out how much space the plants need, how much light they need, what time of year they grow, etc. Then provide a written explanation of how you plan to add plants to the grounds.

25. In many urban areas, people have transformed vacant lots into beautiful and productive vegetable gardens. Imagine that you are an urban gardener, and develop plans to show how you would transform a vacant lot into a garden. Include at least 10 types of plants (vegetable or flowering) in your plan. Do research to find out how much space the plants need, how much light they need, what time of year they grow, etc. Then sketch the area, showing where each plant would go, and provide a written explanation for the logic behind the placement of each plant.

Answers to Unit 8 Extra Assessment Items

Word Usage

1. Answer:
   Sample answer: In order for the plants to survive on Mars, they must live in a *biosphere*. If plants are grown in soil, it should be *loam*. The *percolation rate* of the soil should allow plenty of air and water to reach the plants' roots. If the plants will be grown by *hydroponics*, they will need a solution for *nutrients* and a *rooting medium* for support.

Correction/Completion

2. Answer:
   Sample answer: Insects are needed to pollinate *many* plants *that require cross-pollination. However, some plants are pollinated by other animals or the wind, and some plants are self-pollinating.*

3. Answer: cross-breeding

4. Answer:
   A male plant's flowers have no *pistil, stigma, ovary, style, or egg cells*, while a female plant's flowers have no *stamens, anthers, filament, sperm, or pollen.*

5. Answer:
   Sample answer: . . . it may kill the plant. A concentrated solution surrounding the roots may cause water to leave the plant cells by osmosis, because water concentration inside the plants will be higher than water concentration outside the plants.

Short Essay

6. Answer:
   Sample answer: Pesticides may poison and kill plants or animals that are not pests. The pesticides may harm people who use them, and they can be a danger to pets. Pesticides are also hard to dispose of safely, so they may end up contaminating the water and soil.

7. Answer:
   The answer is acceptable if it mentions that moisture is needed for germination to take place. The answer deserves a bonus if it includes additional advice such as the desirability of planting several seeds (since not every seed will germinate) or of using a sterilized growing medium.

8. Answer:
   Sample procedure: (1) Pinch off petals and anthers. (2) Cover the pistil with a bag. (3) Get anthers from another plant. (4) Using a brush, transfer pollen to the stigma, and cover it with the bag for a short time.

Short Response

9. Answer: Sample answer: Water, air, support, nutrients, space, light, and warmth

10. Answer:
    Controlled factors: plant type, size, and age; pot size; soil; water; light; nutrients; variable: room temperature

11. Answer:
    Possible controlled factors: type of seed, growing medium, light, water, pot size, temperature, number of seeds; variable: amount of moisture in the seed-storage area

12. Answer:
    Possible controlled factors: type of soil, amount of soil, moisture, light temperature, pH testing procedure; variable: amount of compost

13. Answer: Sample answer: air, water, food, light, warmth

14. Answer:
    a. The letters refer to the most important chemical elements that plants need to survive. N stands for nitrogen, P stands for phosphorus, and K stands for potassium.
    b. The numbers refer to the percentage of each chemical present in the fertilizer.
    c. 40 percent

Graphic

15. Answer:
    a. Food production increases.
    b. Food production increases and then decreases slightly.
    c. Answers include temperature, water, nutrients, plant type, soil type, etc.
    d. Answers will vary depending on the variable chosen. All factors would affect the rate of photosynthesis. Too little or too much would decrease the rate. An optimum amount would increase the rate.

16. Answer:
    a. 15 m-30 m per minute
    b. Decrease
    c. About 0.5 cm per week
    d. Growth rate increased when the air flow increased from 0 m to 15 m per minute, and was the highest from 15 m to 30 m per minute. Possible reasons include that air flow provides a fresh supply of carbon dioxide to the leaves, which aids photosynthesis. Air flow also helps to keep the surfaces of leaves dry, which helps to prevent diseases.
    e. Faster air flow causes the growth rate to decrease. Possible reasons include that the air flow causes plants to transpire more water, which dries up or damages the leaves and slows photosynthesis.
    f. Sample answer: A gardener should plant his or her garden in an area that has some air flow but that is protected from high winds (high air flow).

7. Answer:

a.

Day	Plant height (cm)					Average height (cm)
	A	B	C	D	E	
1	0.4	0.5	0.4	0.3	0.5	0.42
2	0.5	0.6	0.6	0.4	0.7	0.56
3	0.7	0.8	0.8	0.6	0.9	0.76
4	0.8	1.0	1.0	0.7	1.2	0.94
5	1.0	1.2	1.3	0.9	1.3	1.14
6	1.3	1.4	1.5	1.1	1.5	*1.36*
7	1.6	1.7	1.8	1.3	1.7	*1.62*
8	1.8	2.0	2.1	1.4	2.1	*1.88*
9	2.1	2.3	2.4	1.5	2.4	*2.14*
10	2.4	2.5	2.6	1.7	2.6	*2.36*

See the graph that follows.

b. See the graph below.

c. The graph indicates that the fertilizer made the radishes grow more quickly.

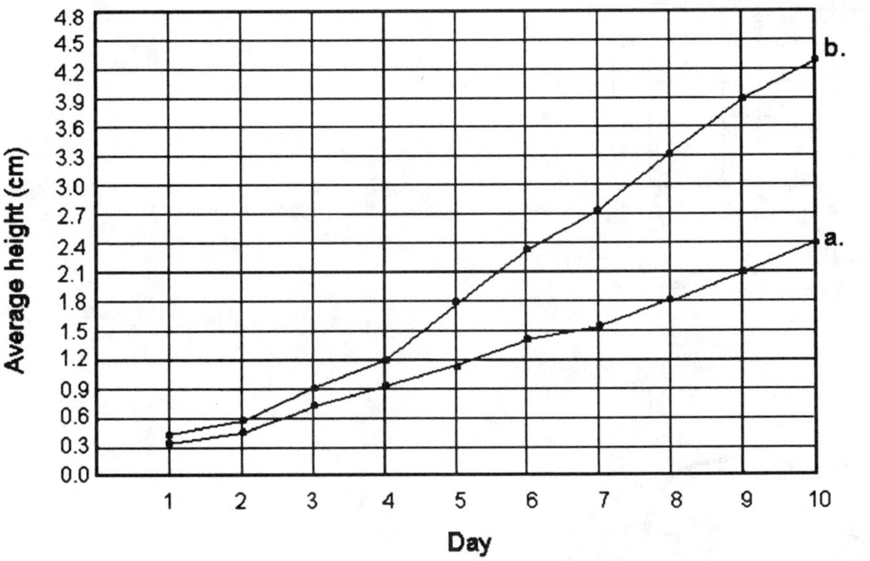

18. Answer:
    a. September, March, April
    b. December, January, June
    c. 3 cm; the soil is closest to the surface and therefore most affected by extremes in the temperature above ground.
    d. 75 cm; the soil at this depth is the most insulated by the soil above it; therefore, it is least affected by the temperature above ground.
    e. The graph for this depth would show less variation because the soil temperature varies less at greater depths.

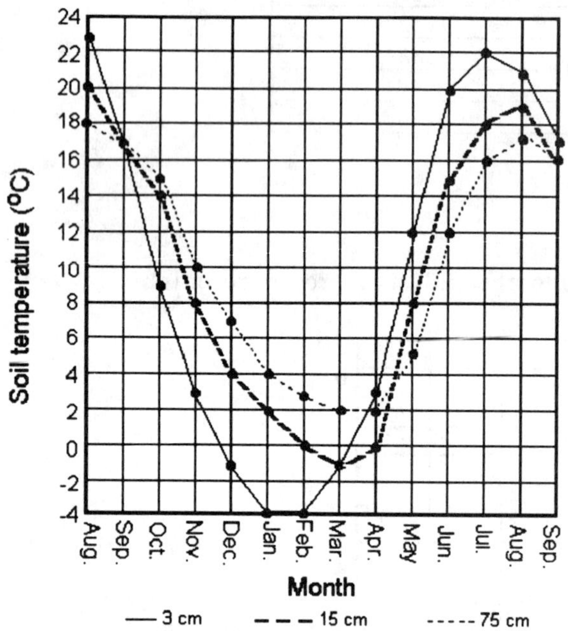

Illustrative

19. Answer:

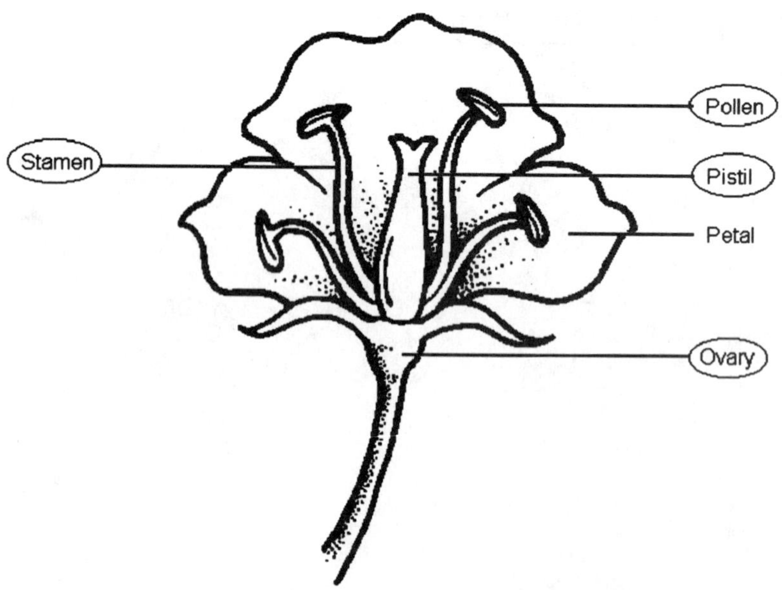

0. Answer:
The answer should describe a method of growing a new plant by using a part of an existing plant. If you are unsure about the answer, compare it with the information in the section Plants From Plant Parts, which begins on page 534 of the textbook. If you believe you could succeed in starting the plant using only the information provided in the answer, give the answer a bonus.

Illustrations should be clearly labeled representations of how to use vegetative reproduction to grow a new plant.

1. Answer:
Plants require light, water, air, nutrients, and a medium for support. The answer is acceptable if the diagram or explanation clearly demonstrates that the plant would obtain all of these without the use of soil.

Illustrations should be clearly labeled representations of how to use hydroponics to grow a plant.

erformance Task

2. Answer:
Answers will vary but should be based on what was learned about organic fertilizers in this unit. Encourage additional research and creative presentations.

3. Answer:
Answers will vary depending on the space chosen, but students should select plants that are logical for the space and should support their choices with clear and accurate rationale.

4. Answer:
Answers will vary but should take into account the factors important for plant growth as well as visual appeal. Sketches should be clear and explanations detailed and accurate.

25. Answer:
Answers will vary but should take into account the factors important for plant growth as well as visual and practical considerations. Sketches should be clear and explanations detailed and accurate.